NIGHTFIGHTER

NIGHTFIGHTER

THE BATTLE FOR THE NIGHT SKIES

Ken Delve

ARMS AND
ARMOUR

Arms & Armour Press
An imprint of the Cassell Group
Wellington House, 125 Strand, London WC2R 0BB

Distributed in the USA by Sterling Publishing Co. Inc.,
387 Park Avenue South, New York, NY 10016-8810

Distributed in Australia by Capricorn Link (Australia)
Pty Ltd, 2/13 Carrington Road, Castle Hill, New South
Wales 2154

British Library Cataloguing-in-Publication data:
A catalogue record for this book is available from the
British Library.

ISBN 1 85409 245 5

Edited by Philip Jarrett

Designed by Roger Chesneau/DAG Publications Ltd

Printed and bound in Great Britain by
Hartnolls Limited, Bodmin, Cornwall

CONTENTS

INTRODUCTION

Any aviator will tell you that night flying is a dangerous occupation, not to be undertaken lightly even with the wealth of equipment fitted to modern military aircraft. Those stages of the sortie when the aircraft is close to the ground, take-off and landing, are particularly hazardous, even on today's well-lit airfields. Having to go and find a target in enemy territory adds yet another dimension of difficulty. So much for the bomber's problems, but what of the defensive fighter trying to find his opponent in the huge three-dimensional space that is the night combat arena? In this age of sophisticated electronic aids we tend to forget the origins of night air warfare, and so fail to appreciate the scale of the problems facing defender and attacker alike.

When the Wright brothers made their first powered, heavier-than-air flights in 1903 they opened a new era of military technology, albeit one that many nations were slow to accept. In Germany the military favoured dirigibles, and the Zeppelins became a mainstay of German air power strategic bombing doctrine as the First World War approached. In these early days, however, it was by no means accepted that night operations would become an essential element of military aviation.

The outbreak of war in 1914 brought rapid advances in technology, and air power became an integral part of most major operations. With the launching of airship raids on England the German air arms opened a strategic bombing campaign that was to have a far-reaching influence. The German press was jubilant, while the British public were alarmed. The political effect of the airship raids far outweighed their military value, and as a consequence the British military had to develop an extensive Home Defence network. The battle for the night skies was truly joined.

The story of the nightfighter in 1915–1918 is very much one of the night war over England, although night operations over the Western Front grew in significance. With the end of the war the victorious Allies set about dismantling their air forces; 1920s doctrine paid little heed to the prospects of offensive or defensive night operations. Thus, at the outbreak of war in 1939, little had changed from 1918. The bomber had trouble finding its target, and the fighter had trouble finding the bomber, although here the scales had begun to tip, owing to the introduction of both ground and air radar. In the first part of the war the nightfighter story is primarily that of the RAF with its development of Airborne Interception, although the 'cat's-eye' fighters played a major role in the early months, countering the growing German night offensive. However, as Bomber Command increased its campaign against German industry and cities, the Luftwaffe had to develop an extensive night defence network, and the emphasis switched rather more towards Germany. Other theatres of the war saw few major developments in nightfighter techniques.

Radar developments were the key to postwar progress, technology increasing the fighter's reliability, accuracy and long-range capability. This, in due course, led to the all-weather day and night interceptor/fighter, the combination of radar and missiles producing an aircraft capable of performing equally well by day or by night. Although night operations have taken place in major post-1945 conflicts such as Korea, Vietnam and the Gulf, the nightfighter as a breed has not really played a prominent role, partly owing to the growth of other elements within an integrated air defence network, the most significant being the surface-to-air missile (SAM). This is not to say, however, that there have not been notable night air actions. As to the future, the advent of stealth technology, exemplified by the F-117, has opened a new chapter in air warfare. These aircraft are primarily night birds, using technology to avoid ground detection and engagement. In the past, such use of the 'big sky' principle was countered by developments in the nightfighter. Will this be the case once more?

Ken Delve

Chapter 1

THE FORMATIVE YEARS

When the Wright Brothers made their historic powered flight in 1903, few could have imagined the developments that would take place over the next fifteen years. By 1918, as a result of the First World War, the aeroplane had become a major factor in warfare.

Progress was at first very slow, with conservative military men and politicians reluctant to adopt the new technology. Much depended on the enthusiasm and support of individuals who could see the aeroplane's military potential. Its primary role was perceived as reconnaissance, and in army manoeuvres it soon became apparent that the aeroplane could play a major part. Even in these early days, however, some saw a more offensive role for the new machines; the ability to fly over the ground forces and drop bombs when and where one chose. In Germany, such a role was identified for the dirigibles (airships) in which that country led the world.

As soon as flying had become an accepted fact, there were bold adventurers who sought to fly by night as well as by day. The earliest such venture was undertaken by Emil Aubrun in a Blériot on 3 March 1910, when he made a 20km round trip from Villalugano, Argentina. Given the low speed of this early aeroplane, the good weather conditions and a well illuminated landing ground (LG), he encountered few problems. However, it was not followed up with any enthusiasm by those who sought military uses for aeroplanes, as it did not fit the primary projected role of reconnaissance. At night the enemy positions and movements could not be observed. Thus, even those military men who saw a role for aeroplanes virtually ignored the night hours.

The Italians were among the leaders in the field of military aviation, making the first military reconnaissance flight in 1911 (over Turkey), developing a variety of roles for their units and employing them on offensive operations in colonial conflicts. Meanwhile, most other 'developed' nations were also introducing aeroplanes into their military forces, usually as an adjunct to, and therefore under control of, the army and, in some cases, the navy. In Britain the Royal Flying Corps (RFC) was formed in 1912 with Military and Naval Wings.

The first night flight by an RFC aeroplane took place on 16 April 1913, when Lieutenant R. Cholmondeley of No. 3 Squadron took his Maurice Farman biplane up on a clear, moonlit night to fly from Larkhill to Upavon and back. He reported no problems on this short flight. The obvious problem with night flying was that it was dark, and ground features were obscured except on very bright moonlit nights. In the absence of such natural illumination, other means were required to allow the pilot to find his destination and land safely. In June 1913 experiments took place using petrol flares at landing grounds, followed over subsequent months by a range of trials using electric lighting, including the installation of two rows of electric lamps along the wing leading edges of two Maurice Farman biplanes. While it is not the author's intention to discuss such developments in any detail, it is important for the reader to realise that without such considerations the art of night flying – and thus night fighting – would have been impossible.

At the outbreak of the First World War in 1914 the fledgling RFC was sent to France to support the British Expeditionary Force, responsibility for the air defence of the United Kingdom falling upon the Admiralty. Throughout 1914 and 1915 this 'air defence' consisted of a small number of guns and searchlights around London, and a variety of Royal Naval Air Service (RNAS) aircraft dispersed along the east coast to patrol the coastline from the Thames to the Humber.

It was already realised that the Germans had the capability of making night attacks using their airships, so the first night defensive patrol was flown on 5/6 September 1914, by an RNAS Farman out of Hendon. Those few units responsible for Home Defence (HD) continued to try their hand at night flying, with very mixed results and a fair number of accidents. Major C. A. Longcroft, the Officer Commanding No. 1 Squadron, expressed the following opinion: 'I see no reason why, with practice – at any rate on moonlit nights – one should not be able to find one's way across country with nearly as much certainty as by day, provided that the destination Landing Ground is made very distinct'. This addressed the basic problem of night operations, but, as yet, no one really addressed the more important question of how to find the enemy in the vast, dark night sky. Despite warnings that the Germans were planning an airship offensive against England, no major effort was made to improve the basic and established aircraft/anti-aircraft gun/searchlight combination.

The first acknowledged British night operation was flown on 21 December 1914 by Commander C. R. Samson RNAS, when he flew a

Maurice Farman in an attack on Ostend. Unable to find the U-boat targets that he sought, he instead bombed coastal batteries. The significance of this night bombing sortie was not truly appreciated by either side.

The British defence system was soon put to the test, and proved to be totally ineffective. Britain suffered its first airship raid on 19 January 1915, when the German naval airships L3 and L4 attacked targets in East Anglia. L3 bombed Yarmouth, killing one man, while L4 flew a course Bacton – Cromer – Sheringham – Beeston – Thornham – Brancaster – Hunstanton – Snettisham – King's Lynn, scattering bombs along the way. Those bombs which fell on King's Lynn killed four people and did £7,740-worth of damage. Although the casualties and damage were small, it was a severe shock to a nation that had previously been beyond the range of European wars. The only defensive reaction was by two Vickers F.B.5s of No. 70 Squadron at Joyce Green, but neither saw the enemy and both force landed; not an auspicious start. This small raid by only two early-model airships was in many ways just a trial while the German airship services (Navy and Army) re-equipped with better machines in preparation for the major effort, the aerial bombardment of London.

In the meantime, however, a number of minor attacks were carried out, including a one-off attack by a seaplane of SFA (See Flieger Abteilung) 1 from Zeebrugge, which bombed Braintree, Essex, on 21/22 February. The outcome of this bold assault was the ditching of the aircraft and the capture of the crew; it was to be many months before aircraft became a threat to the UK. Zeppelin attacks took place on 14/15 April (Tyneside), 15/16 April (Humber area), 29/30 April (East Anglia), 9/10 May (Southend), 16/17 May (Kent coast) and 26/27 May (Southend). Damage and casualties were light, but the defences proved unable to make contact with the raiders, and very few defensive sorties were flown.

However, during the Kent coast raid the defenders did at last manage to get near their target. On this raid LZ38 dropped bombs in the Ramsgate–Dover area in the early hours of 17 May. At one stage the airship was picked up by searchlights, the first such success, and one patrolling aircraft, the Avro 504 flown by Sub-Lieutenant Reford Mulock, spotted the intruder and gave chase. Unfortunately the airship climbed away, and although Mulock chased it all the way to the Belgian coast he did not get close enough for a shot. The Admiralty had warned the continental squadrons, and nine aircraft were sent up from Dunkirk and Furness. None of these found LZ38, but by coincidence they came across LZ39 on its return from an attack on Calais. The Nieuports of Squadron Commander Grey and Flight Sub-Lieutenant Warneford made attacks and claimed hits, but with no apparent result. Some minutes later Flight Lieutenant Bigsworth in an Avro 504 managed to get above the Zeppelin as it neared Ostende. He dropped two 20lb bombs and claimed to have damaged the enemy. It was later discovered that the airship had been hit, one man being killed, several injured, and extensive damage being suffered. This first 'success' for the nightfighters provided a much-needed boost to morale.

London received its first raid on the night of 31 May 1915, when Hauptmann Linnarz in LZ38 dropped incendiary bombs and grenades, killing seven people, injuring 35 and causing £18,596-worth of damage. The other airship on the raid, LZ37, only reached the Kent coast before having to turn back. Casualties and damage were again small, but this attack, seemingly mounted with impunity (fifteen sorties had been flown by the RNAS) by a single airship, held the threat of worse to come, and the terror it caused far outweighed the material damage. The German press was jubilant, one newspaper in Leipzig announcing:

> England no longer an island! The City of London, the heart which pumps the life-blood into the arteries of the brutal huckster [mercenary] nation, has been sown with bombs by German Airships, whose brave pilots had the satisfaction of seeing the dislocated fragments of docks, banks and many other buildings rise up to the dark skies in lurid tongues of flame.

This was strong propaganda, but the fact that England was 'no longer an island' was not lost on the population of London.

The Zeppelins returned three more times during June; 4/5 (London and Humber), 6/7 (London and Hull) and 15/16 (Tyneside). The latter two raids caused heavier casualties and greater levels of damage, while the Home Defence units still failed to achieve anything worthwhile. However, one of the four raiders on the second raid was shot down over Belgium by Sub-Lieutenant Warneford in Morane Parasol 3253. Having bombed Dover, LZ37 flew home via Calais, where a few more bombs were dropped. A this point the airship was engaged by a number of warships. The Zeppelin steered towards Ghent, and at 0150 Warneford made his first attack, only to be driven off by the alert gunners. He circled away, pretending to lose interest, but merely sat a little distance away, biding his time. As the Zeppelin began its descent he moved in above his target and dropped six bombs. Within seconds the air was rent with a major explosion and LZ37 fell to its destruction. Warneford was awarded the Victoria Cross.

That same night a second airship was destroyed just after it had docked at Evère. In one of the first Offensive Counter Air (OCA) missions in aviation history, Henry Farman biplanes of No. 1 Squadron RNAS bombed the base. In later years such attacks on the home bases of night bombers (and fighters) would be an integral element of the night war.

A major raid by five airships on 9/10 August attacked London and the Humber area, resulting in seventeen deaths and £11,992-worth of damage. The RNAS Home Defence units flew 21 sorties, the variety of aircraft involved showing the somewhat ad hoc nature of their equipment. The Royal Aircraft Factory B.E.2c was the primary nightfighter, seven being airborne on this night, along with a Blériot XI, Avro 504B, Sopwith Schneider, Bristol T.B.8, Curtiss S, Short 184, Caudron G.3, and Sopwith Tabloid and two-seater Scout. A mixed bag indeed, and not one of them able to catch a Zeppelin. London was again attacked on 12/13 and 17/18 August.

The public and the press in England continued to express concern at the failures of the defences, especially after an attack by Kapitänleutnant Heinrich Mathy on the night of 7/8 September. Once again damage was slight, but it infuriated the already angry and frightened people of London, who demanded to know why nothing was being done.

Earlier in 1915 Mathy had commented on the value of an aeroplane defence against Zeppelins:

> As to an aeroplane corps for the defence of London, it must be remembered that it takes some time for an aeroplane to screw itself up as high as a Zeppelin and by the time it gets there the airship would be gone. Then too it is most difficult for an aeroplane to land at night.

Despite the notable success of Sub-Lieutenant Warneford, it appears that the British Admiralty agreed with Mathy's assessment, and in September 1915 they concluded that the air defence of London would be best served by improvements in the system of guns and searchlights already in existence, supplemented by a few aircraft. There had been four further raids during September.

For the defenders the statistics so far were not good. Aircraft had flown 89 night defensive sorties for the cost of twenty aircraft wrecked, three pilots killed and eight injured, with not even a hint of success. Discussions continued at the War Office as to who was best suited to perform the task of Home Defence, the Admiralty or the RFC. Neither group wanted the responsibility, and the RFC were claiming lack of equipment, pilots and landing grounds. A compromise was eventually reached, the Admiralty being formally responsible for the task but the RFC agreeing to take a more active part. As a result, on 2 October Lieutenant-General W. G. Salmond, Commanding Officer of No. 5 Wing at Gosport, was ordered to 'bring machines to the vicinity of London for the period October 4–12' to strengthen the air defence network. This reinforcement consisted of one S.E.4a and seven B.E.2c fighters. The landing grounds at Northolt, LGII (later known as Sutton's Farm) and LGIII (later known as Hainault Farm) each received two B.E.2cs, and one B.E.2c and the S.E.4a went to Joyce Green.

Each landing ground was equipped with portable hangars and a number of temporary buildings as offices and working accommodation. Pilots were billeted in the nearest house with a telephone, this being essential as the duty pilot would be alerted by a telephone call from the War Office when a raid was expected. This system was supposed to allow time for the aircraft to climb to 10,000ft and thus give the pilot some chance of catching any Zeppelin he might see. If no enemy was sighted, the pilot was to land 1½hr after take-off. The main problem was the lack of any real warning system, the primary source of such information being intelligence from agents on the Continent and, especially later on, the monitoring of Zeppelin W/T transmissions.

The trial period was officially extended to an 'indefinite period' and a number of attempts were made to intercept the increasing number of Zeppelins now raiding England, although invariably the fighters failed to get within range of their opponents. One attack on 13 October caused such chaos and terror in London, with 71 people killed, that it led to public meeting at Cannon Street Hotel, with people demanding 'a declared policy of air reprisals for Zeppelin raids on London and other open cities and an adequate system of air defence'.

After L15's raid of 13 October 1915, Joachim Breithaupt, one of the crew, wrote:

> It is easy to find the way. London lies beneath us like a lightened map. The Officer of the Watch aims through the pendulum-aiming telescope and releases the bombs. It would be a mistake to suppose that small, isolated targets could be hit. We dropped the bombs where they were certain not to fall into open fields, in parks or into water. The machine-gunners are at their posts. . . Thank God! We have passed the city. But suddenly a fresh murderous fire begins from a direction whence we did not expect it. At the same time enemy aeroplanes dart through the rays of the searchlights, beside us and over us, as we are informed by the lookout on the bridge through the speaking tube. The tracer bullets fired by the pilots are easily seen, the

> slightest contact with which would set our ship on fire. But we are in luck once more! At last after 40 minutes we are away from the 'unfortified' city. Over the coast we are fired at again, but the ill-directed fire does not have any effect; how much better was the marksmanship of the London batteries!

The calls of the people of London fell on deaf ears. Instead of improvements in the air defences, by 26 October all detachments were back with their unit at Gosport, despite the 'indefinite extension'. This meant that there was no aeroplane defence from the recently established landing grounds in and around the city. However, by the end of December, under continued public pressure, the Directorate of Military Aeronautics ordered that two B.E.2cs be stationed at each of ten landing grounds around London, a combination of the previous fields and a number of new ones. The organisation of each field was based on the limited experience gained from the earlier operations, each being allocated six mechanics and a Royal Engineers party with one searchlight and a 13-pounder gun, as well as the two aircraft and their pilots.

Despite this seemingly positive action to solve the problem and subdue the public outcry, there was still no firm policy decision as to who was responsible for air defence. With continued success by the German airship services, and a notable lack of success by the defences, the conclusion was at last reached that the RFC should take responsibility for an aeroplane force for the defence of London. At a War Committee meeting on 10 February, official blessing was given for the formation of specialist squadrons for Home Defence.

At the beginning of April 1916 control of the ten dispersed flights was given to Major T. C. Higgins, 19th Reserve Squadron, which, on 15 April 1916, officially became No. 39 (Home Defence) Squadron, the first of the specialist squadrons tasked with providing an effective aeroplane defence for London against the seemingly untouchable giants of the enemy. The initial establishment of the squadron was two aircraft at each of the ten airfields: Hounslow, Northolt, Hendon, Chingford, Hainault Farm, Sutton's Farm, Joyce Green, Farningham, Croydon and Wimbledon Common. Higgins soon realised that this organisation was too unwieldy to be effective, and decided to concentrate his resources at three fields, Hounslow ('A' Flight), Sutton's Farm ('B' Flight) and Hainault Farm ('C' Flight). Each field had six B.E.2cs plus their support facilities, with Hounslow also acting as the training centre, supply depot and Headquarters. Sutton's Farm and Hainault Farm had been established in mid-1915 during the initial search for landing grounds.

Sutton's Farm, near Hornchurch in Essex, was typical, being a 90-acre piece of farmland, an irregular slab of stubble-land which, by the time No. 39 Squadron took it over, bore some resemblance to an airfield. Nevertheless, it was still only a rough grass strip with many hazards for day flying, never mind night flying. Two portable hangars had been erected, and a number of sheds and tents had been provided for the support facilities of the Flight. Hainault Farm, near Ilford in Essex, was somewhat similar. All three airfields also had a 13-pounder gun and searchlight for 'self-defence', although the searchlights found more use as take-off and landing aids for night flying rather than in engaging enemy raiders.

It was already well proven that the greatest danger to the night-flying aviator lay in trying to get his aircraft back on to the ground. Throughout 1916 various experiments were conducted in an effort to improve matters. Many such experiments took place at RNAS Great Yarmouth, the leading RNAS night defence unit owing to its geographical position in the 'front line' as far as the airship raids were concerned.

One memo stated:

> Certain struts and other parts of aircraft are to be made sufficiently luminous at night by being covered with luminous paint to assist in night flying and to avoid their being accidentally hit when firing at other aircraft. In many cases, in tractor machines, it would be desirable also to paint a small part of the tip of each propeller blade. All machines of which there is the slightest possibility of their being used at night should be so prepared, even although it is not intended to use them so normally or for practice purposes. In addition to the luminous paint, luminous string will, if possible, be supplied in place of the usual string as side-slip indicator. Pending its supply, ordinary string, dipped in luminous paint, must be used.

Regardless of all these aids, it finally fell to the pilot to make a safe approach and landing:

> The peculiar difficulty about night-flying is judging distances. You think you are about 50ft from the ground, when the ground suddenly hits you before you have time to 'flatten out', and as a direct consequence, the machine flies straight into the ground, with disastrous results to the machine.

Only the slow speeds and flimsy structures of the aircraft saved many pilots, the machines being written off but their occupants escaping with minor injuries. Others were not so lucky. Throughout the war very few nightfighters were lost to enemy action, but a great many aircraft were crashed and pilots killed through other causes.

The RFC's primary nightfighter, the B.E.2c, had seen service on the Western Front, where it soon became known as 'Fokker Fodder'. It had been designed as a simple, easy-to-fly aircraft, which is exactly what it proved to be, but it was far too stable for the type of air warfare that had developed in Europe. However, this inherent stability proved to be a great advantage to No. 39 Squadron, as the aeroplane was an excellent gun platform for night warfare and was easy to handle. It also had a number of disadvantages, having a top speed of only 72mph with its 90hp RAF engine, and taking 45min to reach its ceiling of about 10,000ft. The B.E.2c's performance was vastly inferior to that of the latest Zeppelins, which, even when they were intercepted, were able to drop ballast and quickly climb away from danger, having a much higher ceiling than the fighter.

Because of these disadvantages it was essential for the fighter pilots to have sufficient warning of a raider to enable them to take off and climb to a reasonable height, where they would have some chance of making an interception. As a partial solution to this problem the duty pilot system was devised, whereby the duty pilot slept next to the telephone and his aircraft was kept at readiness outside the hut. The ground crew would start the engine at regular intervals to make sure there would be no delay when the warning came. Standing Orders for No. 39 Squadron required that the duty aircraft be airborne within five minutes of receiving the warning, and should then take position in the system of patrol lines. Designated patrol lines and heights were allocated to No. 39 Squadron and the other four Home Defence squadrons which had been formed by mid-1916 (No. 33 Squadron at Bramham Moor and Knavesmere in March, No. 36 at Cramlington, No. 50 at Dover and No. 51 at Norwich, all formed in May). Patrolling aircraft had to fly above 8,000ft, the height of the London balloon cordon, and maintain station by reference to any ground lights visible.

Identification of raiders would usually depend on their being illuminated by the searchlights. Often the first indication was the sound of the anti-aircraft guns. This attracted the attention of the nightfighter pilots, who would then scan the appropriate sector of the sky, hoping to spot a Zeppelin in a searchlight beam. The point at which they chose to fly towards the area was decided by each pilot; Napoleon would have approved, as this fitted his maxim of 'marching in the direction of the sound of the guns'.

Maintaining position on the dark nights usually favoured by raiders was not always easy, as a Norwegian pilot on No. 39 Squadron discovered. During 1916 the Norwegian Government became increasingly concerned over the threat to their cities posed by the Zeppelins, and a Norwegian officer, Tryggve Gran, was seconded for service with the RFC, joining No. 39 Squadron in December 1916. In the following account Gran describes a typical call-out:

> The rain was pouring down in torrents and it whistled in the stove pipe as if it was blowing a hurricane outside. We were discussing the Norwegians and the part they had played in the war when suddenly the alarm went. I felt the blood rush to my head as I ran outside – Great Scott, what a night – rain and hail hit me in the face and dark clouds were passing just above the tree tops.
>
> Captain Stammers had gone to the telephone and after taking a message he came back. 'Zepps, my boy – stand by first machine,' he shouted. The ground flares were lit and two searchlights endeavoured in vain to penetrate the dark drifting clouds. Like drowned rats the mechanics ran round the machine getting it ready.
>
> 'You cannot fly tonight, sir, it's impossible,' said my fitter. Very seldom in my life have I felt my courage pass away as on this dark stormy night.
>
> 'Come down again if it's too bad,' Stammers said. I gritted my teeth together, waved the mechanic away and let my machine rip out into the darkness. I saw the last landing flares disappear under my wings, and I felt the rain and hail beating my face. Then suddenly everything turned into a chaos of fog and darkness in which only my instruments could be seen. My plane was terribly chucked about and for a moment I completely lost control.
>
> Suddenly I discovered under the brim of my upper plane some flaming flares – I was flying upside down and with a terrible speed. With a purely mechanical reaction I did a 'half-roll' and came on to an even keel. Putting the nose of my machine into wind I determined to hold that course until clear of cloud.
>
> The following 3/4 hour seemed to me to be like years, and I sat looking only at my compass and instruments. Then it became lighter, the motor seemed to get an easier breath and before I knew what had happened it was quite clear with my plane passing over a huge white ocean of cloud. It was like coming into the land of the fairies. For a moment I was sitting in bewilderment at this wonderful splendour of nature.
>
> Then I remembered my orders to patrol the line between North Weald and London Colney at 12,000ft, for three hours. This was easier said than done! An hour I had been airborne without seeing anything and I had only a vague idea of my position. My aneroid was showing 13,000ft and I presumed that at this height the wind would be blowing almost due west. Lower down the wind was south-westerly and I had been steering that course for 3/4 hour. Giving the speed of my machine to be 60mph, and a headwind of 30mph, then I should be some 20 miles SSW of the airfield. Consequently I turned my machine on to north-east and kept going for a little over 1/4 hour, hoping to see some sign of the airfield – but there was nothing to see except the same ocean of cloud.

> Of the Zeppelins there was no sign, nor of any other aeroplane, and as the minutes passed I started feeling rather lonely. It was cold too, and I could see the ice crystals glistening in the moonlight on my stays and wings.
>
> Two hours I kept going backwards and forwards and the faint colour of the day started spreading over the horizon. Then everything went quiet . . . the blood rushed to my head and my heart beat violently; the engine had stopped and nothing would make it start again. Downwards towards the layer of clouds I went, whilst the pressure of the wind whistled in the stays and rigging. Everything turned to cloud an fog. I arranged my landing flares and turned the aeroplane into the presumed wind. The aneroid was showing 2,000ft and still nothing but fog – I dropped a parachute flare which fell open and lit, revealing nothing but fog. Carefully, in huge circles I followed the falling star. Then suddenly the clouds disappeared and I saw the earth – a forest, a river, a few roads. I turned my aeroplane and noticed a field within easy reach, I lit the two magnesium flares under my wingtips and glided into the grassy field.

This excellent account highlights the many and various problems suffered by the nightfighters. Gran had in fact landed near Hull!

The raiders were still able to wander freely over England, seemingly immune from aircraft and guns. Liverpool was the target of a major attack in the late hours of 31 January 1916, nine Zeppelins being involved. The RFC was now beginning to take a more active part in Home Defence activities, and flew 14 of the 22 sorties that night, all but one of the aircraft involved being B.E.2cs (in addition there were 10 sorties by the Nieuports from Dunkirk). Casualties and damage were heavy, and the nightfighters had a bad night, at least six aircraft being wrecked in crashes and two pilots killed.

The attacks continued throughout the spring. April was a particularly hectic month, operations taking place on seven nights. Although the defenders recorded more sightings, and even a few attacks, there were no successes.

Airship raids still predominated, although a few aeroplane attacks also took place, the largest being by six aircraft against the Kent coast on 19/20 May. The eight defensive sorties were airborne far too late.

For the remaining months of the summer of 1916 the public continued to demand action against the German raiders, who still seemed to attack with impunity. Meanwhile, the pilots of the Home Defence squadrons were continuing to gain experience in the hazardous occupation of night warfare and anti-Zeppelin operations. Certain flight commanders, including Lieutenant W. Leefe Robinson at Sutton's Farm, instituted studies of Zeppelins and their performance, and examined accounts of previous engagements to try to discover a means of defeating the giants. This led to unofficial modifications of aircraft and armament that became characteristic of HD units.

August was another busy month, with five raids. The only significant development by the defenders was the employment of an 'aircraft carrier' on the night of 2/3 August. HMS *Vindex* sailed from Harwich with Bristol Scout D number 8953 and its pilot, Flight Lieutenant Charles Freeman. The intention was to spot the enemy further from the coast, and thus catch him unawares while he was lower down. Airborne at 1950, Freeman soon saw two airships and gave chase. Despite a number of technical problems he made a couple of unsuccessful attacks before engine failure forced him to ditch. Freeman was eventually rescued by a Belgian ship and made his way back to England. Also airborne that night was Flight Lieutenant C. J. Galpin in Sopwith Schneider 3776 from Great Yarmouth, who was among the few pilots airborne that night to sight the enemy:

> I left Yarmouth at 7.15pm, and chased the Zeppelin at 6,000ft for about 25 miles northward without bringing her into action. The light and my petrol both failed at 8.45pm, but I was fortunate to descry a ship and alight near her before my petrol petered out. I put the machine down at the bow of the vessel and, as I drifted past, shouted for them to throw a rope. . . . Unable to tow the seaplane against the tide, those fellows stood by me in an open boat from 9 o'clock to 4 o'clock in the morning, when they managed to get me and the seaplane in board with so little damage that after arriving at Grimsby on the following day, I was able to take off and return to Yarmouth.

Once again the Zeppelins had escaped without loss or damage, and in most cases their crews were not even aware that the fighters had been anywhere near.

Saturday 2 September 1916 was a dull, rainy day, and no flying was scheduled at Sutton's Farm, the comment being made that 'it did not look like Zeppelin weather today'. The opposite proved to be the case, and a large raid of sixteen army and navy airships began crossing the East Coast at about 2200.

At 2305 the telephone rang in the duty office at Sutton's Farm. 'Take air raid action!' shouted the operations officer. Robinson was duty pilot, and he climbed into his B.E.2c, 2092, and checked his cockpit lighting and his single Lewis machine-gun, with its ammunition drums full of the new Brock and Pomeroy incendiary bullets, plus tracer for sighting. Satisfied, he settled into his seat and gazed at the cloud-covered night sky.

'Petrol switches on.'

'Suck in.'

Thumbs up . . . 'Contact.'

The engine burst into life and the airfield flares were lit as Robinson yelled 'Chocks away!' and 2092 bumped off down the field and took off. Robinson soon climbed above the shallow ground fog into the clear September sky. His orders were to position himself on the patrol line between Sutton's Farm and Joyce Green. At this point the words of his official report, preserved in the Imperial War Museum archives, take up the story.

> I went up at 1108 pm with instructions to patrol between Sutton's Farm and Joyce Green. I climbed to 10,000ft in 53 minutes. I counted what I thought were sets of flares – there were a few clouds below me but on the whole it was a beautifully clear night.
>
> I saw nothing until 1.10 am when two searchlights picked out a Zeppelin southwest of Woolwich. The clouds had collected in this quarter and the searchlights had some difficulty in keeping up with the aircraft; by this time I had managed to climb to 12,900ft and I made in the direction of the Zeppelin, which was being fired on by a few anti-aircraft guns, hoping to cut it off on its way eastwards. I very slowly gained on it for about ten minutes – I judged it to be about 200ft below me and I sacrificed my speed in order to keep my height. It went behind some clouds, avoided the searchlights and I lost sight of it. After about 15 minutes of fruitless search I returned to my patrol. I managed to pick up and distinguish my flares again. At about 1.50 am I noticed a red glow in the north-east of London. Taking it to be an outbreak of fire I went in that direction. At about 2.05 am a Zeppelin was picked up over NNE London (as far as I could judge).
>
> Remembering my last failure, I sacrificed height (I was still at 12,000ft) for speed and made nose-down for the Zeppelin. I saw shells bursting and night-tracer shells flying around it. When I drew closer I noticed that the anti-aircraft fire was too high or too low, also a good many rose 800ft behind. I could hear the bursts when about 3,000ft from the Zeppelin.
>
> I flew to about 800ft below it and fired one drum along it from bow to stern. It seemed to have no effect. I therefore moved to one side and gave it another drum along its side – without much apparent effect. I then got behind it (by this time I was very close – 500ft or less below) and concentrated one whole drum on one part underneath. I was at a height of 11,500ft when attacking the Zeppelin. I had hardly finished the drum when I saw the part fired at glow. In a few seconds the whole of the rear part was blazing. When the third drum was fired there were no searchlights on the Zeppelin and no AA was firing. I quickly got out of the way of the falling Zeppelin and, being very excited, fired off a few red Very lights and dropped a parachute flare.
>
> Having little oil or petrol left, I returned to Sutton's Farm, landing at 2.45 am. On landing I found that I had shot away my machine-gun wire guard,

the rear part of the centre section, and had pierced the main spar several times.

The flaming wreck of the Schütte-Lanz SL11 (which had a principally wooden structure, unlike the metal-framed Zeppelins) crashed at Cuffley in Hertfordshire. Leefe Robinson was exhausted, stiff and frozen when he landed, to receive a telephone message from headquarters congratulating him on his success and requesting his report. He collapsed on his camp bed and fell into a deep sleep, only to be roused early on the Sunday morning by jubilant fellow officers wanting to drag him off to Cuffley to see 'his' airship. By the time they reached the site of the burnt-out wreck it was already crowded with sightseers; Robinson's combat and the flaming destruction of the SL11 had been witnessed by millions in and around London. The site was guarded by RFC and Army personnel in an effort to preserve as much of the wreck as possible for intelligence purposes. It took the party from Sutton's Farm some time to persuade the guards that they had some claim to a quick look. Within 48hr more than 10,000 spectators had been to the site and numerous pieces of the SL11 had been removed as souvenirs. Hauptmann Wilhelm Schramm and his fifteen crew members all perished in the crash, and were later given a full military funeral; much to the annoyance of many Londoners, who considered them no better than murderers. Leefe Robinson's opinion is not recorded.

It had been the largest raid to date, with fifteen airships taking part, intended to inflict major damage on London. At long last the defenders had achieved something.

Any image of the lone fighter searching the night skies for the intruder would be mistaken, as Robinson was only one of many airmen airborne that night. The first patrols by No. 39 Squadron were sent up at about 2300, while the second patrol, which was airborne at about 0100, consisted of Second Lieutenant F. Sowrey from Sutton's Farm, to patrol Joyce Green to Farmingham; Second Lieutenant J. I. Mackay from North Weald Bassett covering the North Weald to Joyce Green line; and Second Lieutenant B. H. Hunt from Hainault Farm to patrol the line Joyce Green to Farmingham. Ross remained airborne for two hours and then crashed his machine on landing, a frequent occurrence in these early days of night flying. Sowrey came back with engine trouble after a quarter of an hour, and Brandon after half an hour.

Mackay, however, reached his patrol height of 10,000ft and, when near the Joyce Green end of his patrol line, he saw the SL11 held by a searchlight to the north of London. Giving chase, he was within a mile of the airship when it suddenly burst into flames and tumbled earthwards. Returning to Joyce Green, he spotted another airship over towards Hainault and gave chase, only to lose sight of it after about fifteen minutes. A similar situation confronted Hunt, who was just about to attack the SL11 when it burst into flames. In the sudden brilliant light he spotted the L16 only a short distance away. He at once pursued it, but, dazzled by the glare, he lost sight of the airship and resumed his patrol line, only to have a third frustrating chase about half an hour later. Other units were active in addition to No. 39 Squadron; No. 50 Squadron at Dover sent up three machines, and aircraft were airborne from No. 33 Squadron at Beverley and from various Naval Air Stations.

The initial newspaper reports made no mention of Leefe Robinson, but everyone was united in the jubilation over this victory against the previously immune raiding giants. It was not until 5 September that the press announced the pilot's name, and the public then insisted that the 'hero' should be rewarded. This was soon followed by the official announcement of the award of the Victoria Cross to Second Lieutenant Leefe Robinson for his action. He was the only nightfighter pilot to win this award. His photograph appeared everywhere, and he became the darling of the nation, particular affection being shown by the people of London. Monetary rewards totalling some £3,500 had been offered by businessmen for the first airman to shoot down a Zeppelin over Britain, and these now went to Leefe Robinson. There was some debate as to whether, as a 'gentleman', he should accept such a reward. The War Office soon solved the problem by passing a regulation preventing any such public acknowledgement in the future.

The Naval Intelligence Division at Whitehall had by now mastered interception of the radio messages passed by the Zeppelin crews as they were forming up for raids, and so were able to give the ground and air defences some forewarning. The system was to prove its worth against a raid on the night of 23 September 1916. This raid, by four of the new Super Zeppelins and eight older models, included Kapitänleutnant Alois Böcker in L33 and Oberleutnant Werner Peterson in L32, two experienced Zeppelin commanders. After crossing the east coast, Böcker steered a direct course for London and was soon engaged by the searchlights and guns surrounding the City, which, for once, were proving to be very accurate. L33 was bracketed by bursting shells and, near Bromley, one shell burst inside the airship's hull, destroying one gas cell and riddling others with shrapnel. Although the hydrogen in the cells did not ignite, it was rapidly escaping from the ruptured cells into the atmosphere. Böcker turned his vessel towards the sea while his crew tried to repair the damage and reduce the rate of descent, which by now had reached some 800ft/min.

By this time the aircraft defences were also alert and active and a number of B.E.2s were airborne, including one flown by Second Lieutenant Alfred de Bathe Brandon, an experienced anti-Zeppelin pilot who had joined No. 39 Squadron when it formed in April 1916. Brandon sighted the crippled L33 over Chelmsford, Essex, and attacked her for some twenty minutes, firing Brock and Pomeroy ammunition with no apparent effect until he lost the airship in cloud. Meanwhile, Böcker and his crew were desperately trying to maintain height by throwing overboard as much weight as they could. This availed them little, and the stricken ship continued to sink earthwards, coming to rest in the marshland of Mersea Island. The crew jumped from the airship, and Böcker destroyed it with a few well placed signal flares. Shortly afterwards he and his crew were taken prisoner by the local policeman, who arrived by bicycle, attracted by the blazing wreckage.

There has always been some doubt regarding the part played by Brandon's attack in the destruction of L33, and whether the ship would have been doomed without his intervention. He must have caused further damage to the gas cells, although members of the airship's crew later stated that they were unaware of his attack. Rightly, he was officially credited with a hand in the destruction of L33.

Later the same night, Brandon spotted another Zeppelin caught in searchlight beams and headed towards it. This was Peterson's L32, and it had already been spotted by Lieutenant Fred Sowrey in B.E.2c 4112. Sowrey, another experienced campaigner on No. 39 Squadron, had, like Brandon, been ordered up on the instruction to take air raid action. He had been patrolling for some two hours when, at 0045, he noticed an airship caught in the beams of the searchlights. In the words of his report:

> I at once made in this direction and manoeuvred into a position underneath [the Zeppelin]. The airship was well lighted by searchlights but there was no sign of any gunfire. I could distinctly see the propellers revolving and the airship was manoeuvring to avoid the searchlight beams. I fired at it. The first two drums of ammunition had apparently no effect, but the third one caused the envelope to catch fire in several places, in the centre and on the front. All fire was traversing fire along the envelope. The drums were loaded with a mixture of Brock, Pomeroy and tracer ammunition. I watched the burning airship strike the ground and then proceeded to find my (landing) flares. I landed at Sutton's Farm at 1.40 am on the 24th. My machine was B.E.2c 4112. After seeing the Zeppelin had caught fire, I fired a red Very light.

Like all pilots who flew at 10,000ft and above in an open cockpit, Sowrey was suffering from the effects of cold and lack of oxygen. Leefe Robinson was one of the first to congratulate him, and concocted a warm drink to revive the exhausted pilot. As soon as Sowrey had scribbled his report he was bundled into Leefe Robinson's new Prince Henry Vauxhall car (part of the 'proceeds' from the SL11 reward) and driven off to the scene of his success. The destruction of L32 was again witnessed by the population of London, and, as in the case of SL11, the site was soon swarming with sightseers. The wreckage fell at Snail's Hill Farm, South Green, near Billericay in Essex, where it burned for 45min. A correspondent for The Times reported:

> [The airship] lay with her nose crumpled and bent out of shape, but the framework of her girders was strong enough to hold together. As she lay it did not seem possible that the fabric was burnt off its gaunt ribs until one noticed pieces of molten aluminium and brass in the debris.
>
> One realised the cost of such a craft even looking at the wreck. Lying on the ground was a red leather cushion. This covered the seat of the engineman, and the ghastly evidence still to be seen showed that he died in his post. One at least of the petrol tanks had burst in half, and the heat of the burning spirit had melted the edges until they looked like some fine fretted lace. There were the remains of an air mattress and a blanket. Curious evidence of the crews' breakfast still remained. There were slices of bacon and hunks of brown, greasy 'Kriegsbröd' with delicately sliced potatoes.

A grim picture of the reality of the total destruction of a Zeppelin. Naval Intelligence officers also discovered a copy of the latest codes, an invaluable aid for the radio interception service.

For their actions on the night of 23 September, Brandon and Sowrey were both awarded the Distinguished Service Order (DSO). Brandon's name never became well known by the general public, as his victory was not in the spectacular fashion of Robinson's or Sowrey's. The score to No. 39 Squadron was now three Zeppelins in three weeks. Morale on the squadron rose to an even greater height, and the exploits of the Zeppelin destroyers became popular throughout Britain. The 'man in the street's' admiration for No. 39 Squadron was immense, and even the inhabitants of the villages around the airfields displayed a new level of tolerance of the pilots' antics. Other HD units were equally jubilant, although it seemed 'unfair' that one squadron was getting all the action. However, the previously low level of morale now soared.

In contrast, morale among the German airship services was no longer so high, and the loss of two naval airships in a single night was to have far-reaching consequences. These victories established the credibility of the aeroplane defences, contrary to the opinion expressed by Mathy some years previously. There was to be only one more large-scale, determined attack on London by the airships, the task of attacking the enemy capital later falling to German strategic bombers.

On the night of 1 October a force of Zeppelins crossed the east coast on course for London. At midnight, Robert Koch in L24 reported seeing an airship ablaze and tumbling to earth somewhere north-east of the docks. It was sister-ship L31, falling victim to another pilot of No. 39 Squadron. Second Lieutenant Wulstan Tempest had been ordered up some time earlier and, having one of the stripped-down B.E.2s, climbed through his designated height of 8,000ft to the then impressive altitude of 14,500ft. At about 2345 he spotted an airship caught by searchlights and turned towards the scene:

> As I drew up to the Zeppelin, to my relief I was quite free of AA fire for the nearest shells were bursting quite 3 miles away. The Zeppelin was now nearly 12,700ft high and climbing rapidly. I therefore started to dive at her, for, though I held a slight advantage in speed she was climbing like a rocket and leaving me standing. I accordingly gave a tremendous pump at my petrol tank and dived straight at her, firing a burst into her as I came. I let her have another burst as I passed under her and then, banking my machine over, sat under her tail, and flying along underneath her, pumped lead into her for all I was worth. I could see tracer bullets flying from her in all directions, but I was too close under her for them to concentrate on me. As I was firing I noticed her begin to go red inside like an enormous Chinese lantern and then a flame shot out of the front part of her and I realised she was on fire. She then shot up 200ft, paused, and came roaring straight down on me before I had time to get out of the way. I nose-dived for all I was worth, with the Zepp tearing after me, and expected every minute to be engulfed in flames. I put my machine into a spin and just managed to corkscrew out of the way as she shot past me, roaring like a furnace. I righted my machine and watched her hit the ground with a shower of sparks. I then proceeded to fire off dozens of green Very lights in the exuberance of my feelings.
>
> I glanced at my watch and I saw it was about ten minutes past twelve, I then commenced to feel very sick, giddy and exhausted, and had considerable difficulty in finding my way to the ground through fog, and in landing I crashed and cut my head on my machine gun.

The L31 crashed in a field just outside Potters' Bar, Hertfordshire, killing all of her crew. Second Lieutenant Tempest was added to No. 39 Squadron's roll of honour as a Zeppelin destroyer, and for his exploit was awarded the DSO.

Thus, in one month, No. 39 Squadron had been responsible for the destruction of four of the 'invincible' monsters. There was only one more raid in 1916. Ten airships attacked the Midlands on 27/28 November, to be met by 40 defensive sorties which resulted in another series of blows to the Zeppelin force. Just before midnight, Second Lieutenant Ian Pyott of No. 36 Squadron was patrolling in B.E.2c 2738 when he saw L34 caught by the Castle Eden searchlights. Diving to attack, he flew alongside the airship and poured machine-gun fire into its port side. The tracers entered the vessel and the pilot concentrated his fire on one spot, this having been shown to be the most effective technique. Soon the flames took hold, and the Zeppelin fell into the sea off the Tees. The L21 was also attacked, but with no apparent result, until it was found again just before dawn by two RNAS pilots in the Yarmouth area. The B.E.2cs of Edward Pulling and Egbert Cadbury made concerted attacks and the airship fell into the sea.

Home Defence had ended 1916 on a high note. The Zeppelin menace seemed to have been beaten, but could they maintain their success in the face of technical advances made by the enemy?

The situation in the early months of 1917 was little changed from that of the latter part of the previous year. The strength of HD units stood at twelve squadrons, most operating from airfields with at least some night flying facilities. However, there was a constant struggle to obtain equipment, and pleas for better aircraft were met by counter-pleas of the needs of the Western Front. Moreover, many weeks had elapsed since the Zeppelins had appeared over England, and many believed that this particular battle was won and that it was ridiculous to have so many squadrons, and trained aircrew, doing nothing while the situation regarding the air war in France was still so desperate. One immediate result of this was the posting of experienced crews to form new night bomber units equipped with the Royal Aircraft Factory F.E.2b. The initial phases of this reorganisation were carried out under the auspices of No. 51 Squadron. Thus, although the HD squadrons had a strength of 222 aircraft by March, only half of these were considered 'available', and almost all were the same tired, limited aircraft with which the defences had been equipped since 1916. The major problem, however, concerned pilots. Only 82 were available, and by no means all of them could be considered experienced in nightfighter operations.

The continuing threat from the Zeppelins led to a number of specialist nightfighter aircraft designs being put forward to the RFC. The Royal Aircraft Factory N.E.1 was a pusher biplane of which six are believed to have been built, although trials at Orfordness did not show much promise and the type did not enter squadron service. Most other designs either never left the drawing board or were limited to one or two prototypes. Such was the case with the Robey-Peters Davis Gun Carrier, the Supermarine P.B.31E, the Parnall Scout and the Vickers F.B.25, all of which were abandoned early in their development. In the absence of replacements, the B.E.2c and F.E.2b remained the RFC's main nightfighters.

For many months the Germans had been working on a range of technical improvements which would restore the invulnerability of the Zeppelins. After a three-month lapse, a five-airship raid was launched on the night of 16/17 March. The target was London, and they came across southern England at about 18,000ft. Although the defenders flew seventeen sorties, they had no hope of reaching the attackers, and it was only bad weather that prevented any concentrated bombing, although the jamming of certain navigational frequencies also played a part, preventing the raiders from obtaining navigation fixes.

By 1917 aerial activity over the Western Front was intense. Reconnaissance machines were now protected by mass formations of 'scouts' intent on destroying their counterparts. Bombing raids become a regular feature of operations, although the tactical element of supporting ground troops was never developed on any appreciable scale during this conflict. However, the strategic employment of bombers was certainly an accepted part of the aeroplane's role, and both sides undertook such missions against a wide range of targets behind the front line, communications targets such as railway stations frequently receiving attention. Partly as a result of the growing danger of operating by day, but also in response to improved conditions for night operations (i.e. better airfield facilities), night bombing was increasingly common. Hitherto, such raids had been confined to the Continent, but on the night of 6/7 May a lone German Albatros C.VII made a daring attack on London, dropping a number of small bombs. The raid appears to have been instigated by the unit commander, without sanction from higher authority. Damage was light and there was only one casualty, but the aeroplane passed almost unnoticed by the defences, four sorties being flown. This intrusion caused no concern among the defenders, but it was a portent of what was to come.

In the meantime, Germany's strategic offensive remained with the airships, and three more attacks were launched during the summer of 1917. London was the target on 23/24 May and 16/17 June, but both raids were badly affected by adverse weather conditions and were largely ineffectual. During the June operation the Germans lost L48, their newest and best airship, shot down by an F.E.2b from Orfordness. This was yet another blow to the attackers, who had hoped that these new vessels would be invulnerable. In fact, L48 was suffering engine trouble and it was this that enabled the F.E.2b pilot to make his attack. It was not the only loss that month. On the night of 13/14, L43 had left Nordholz to patrol the U-boat blockade area, other Zeppelins being airborne the same night on similar naval tasks. Radio messages were intercepted in Britain and direction-finding positions obtained. At 0515 Curtiss H-12 flying boat 8677 took off from Felixstowe and headed towards the area. While patrolling near Vlieland at about 0840 the crew saw and engaged the L43 with machine guns. The attack was effective and the Zeppelin fell in flames, the first success for the flying-boat units.

The final raid of this series took place on 21/22 August and was aimed at the north of England. Eight airships took part, and although the defenders flew 21 sorties they were unable to reach the Zeppelins' operating altitudes. Thus, in many respects, the contest seemed evenly balanced. It was now to take a more dramatic turn.

On 2/3 September No. 70 Squadron sent up a number of its Sopwith Camels from Poperinghe in an attempt to intercept a German night bombing raid directed at St Omer, and Bristol Fighters of No. 48 Squadron were airborne from Leffrinckhoucke after raiders in the Dunkirk area. That same night saw the first German aeroplane night raid on England, a number of aircraft, probably from Kaghol 4, attacking Dover just before midnight. Two bombers cruised over their target at 2,500ft and caused a certain amount of damage and one fatality. The operation caught the defences off guard and no sorties were mounted, though a Sopwith Pup and an Airco D.H.5 were airborne from Dover on training sorties and patrol duties.

A year earlier the German Riesenflugzeug (giant aircraft, or R-planes) had started their operational career on the Eastern Front, and soon had to seek the cover of darkness to escape the Russian defences (the first night operation took place in August 1916). Most of the early sorties met with few problems apart from certain technical difficulties with the aircraft, though crews reported seeing Russian aircraft from time to time. On the night of 16 August the seaplane base at Lebara was raided by the Staaken RML.1, and while the bomber was overhead its crew observed four seaplanes taking off. Two gave pursuit, one getting close enough to make an ineffective attack on the RML.1. Before it was driven off by defensive fire it had scored a few minor hits on the bomber. Gradually the first Riesenflugzeug Abteilung (giant aircraft unit), Rfa.500, received more aircraft and began to increase the scale of its operations, joined by Rfa.501.

However, it was on the Western Front and in the strategic bombing of England that the R-planes were to be primarily employed. The strength of R- and G-type heavy bomber units was increased, and the first raids, by day, were launched during May to August 1917. Losses forced a change to night attacks. In the meantime the experienced Rfa.501 had moved to the Ghent area, coming under the operational control of the Gotha-equipped Bombengeschwader 3.

The initial attack was followed up on the next night, 3/4 September, by a raid against Chatham, Kent, by five Gothas of Kaghol 3. One of the bombers turned back with technical problems, but the others found their target. Although only twenty or so bombs fell near the Chatham docks area, a number struck a barrack block, killing 130 naval recruits. The RFC flew sixteen defensive sorties, some of which were on an exercise, but no contacts were made. However, the most significant move from the defenders' point of view was that Captain Gilbert Murlis Green, the CO of No. 44 Squadron at Hainault in Essex, had been given permission to despatch three of his Sopwith Camels. It had previously been considered that night operations by these single-seat fighters were impracticable, although the pilots had been agitating for some while to be allowed to try. It was certainly time for an aircraft with better performance to join the hard-pressed B.E.2 and F.E.2b stalwarts.

The bombers returned on the night of the 4/5th. The target this time for the eleven Gothas was London. Only five reached and bombed the target, causing casualties and damage; others bombed various coastal targets. The defences were active, eighteen sorties being flown. Flying an Armstrong Whitworth F.K.8 of No. 50 Squadron, Second Lieutenants Grace and Murray spotted the exhausts of an aircraft and, receiving no reply to their recognition signals, attacked. After getting in a couple of bursts of fire they lost the other aircraft, not having observed any result. One other defender claimed to have attacked a bomber, again with no result. However, the raiders lost one of their number to anti-aircraft fire.

Having become used to the airships, the nightfighter crews found the bombers much harder to detect. Searchlight beams or the explosion of the raiders' bombs remained the main indicators of the enemy's presence. Even when they had acquired an enemy aircraft, which was often betrayed by its exhaust glow, the fighter crews found it hard to assess range at night to ensure that their restricted chance of engagement was successful. Unfortunately there was no substitute for experience, and that could only be gained over a long period of time.

The next major attack came three weeks later, when sixteen Gothas set out from their bases on the night of the 24th with London as their target. However, for a variety of reasons only nine made attacks against targets in England, a few hitting London but others bombing Dover, Margate and various locations in East Anglia. But the events of the night of 24/25 September were not over, as the Germans had planned a double blow, with a major airship raid following the bomber attack. Eleven airships were sent to targets in the Midlands and northeast England. It was a bold plan, but once again the airships had a bad night. Only five appear to have reached the target areas, and they caused very little damage. The defenders mounted 36 sorties but no combats took place. Although a number of other airship raids were made before the end of the war, such attacks were proving increasingly ineffective and costly.

The night war had taken a new turn. No longer was it a 'simple' matter of finding cumbersome airships as they drifted across the searchlights, it was now a 'needle-in-the-haystack' situation of trying to find an aeroplane. Among the many panic comments was one by General Smuts: 'Our aeroplanes afford no means of defence at night as they find it impossible to see the enemy machines even at a distance of a couple of hundred yards. They are at night useful only against very large and conspicuous objects, like Zeppelins.' That being the case, many proposed that reliance would have to be placed on the guns (and therefore on a requirement for Home Defence to have a priority in this respect), and the new devices such as balloon barrages. Trials with the latter had taken place in September and the first operational sections were in place the following month.

Bombers returned again on 25/26 and 28/29 September, the latter being planned as the largest raid to date, with 25 Gothas and, for the first time, two giants. London was once more the target, but bad weather ruined the plan and none of the raiders reached the city. Recovering to their home bases in bad weather, six of the Gothas crashed.

It was a similar story on the night of 29/30th, with a Gotha/Giant force aiming for London but being dispersed by bad weather. Two interesting developments were made by the defenders. Firstly, the trials unit at Orfordness flew a Martinsyde F.1 fitted with a pair of Lewis guns mounted to fire upwards (shades of a device that would wreak havoc among RAF bombers in 1944); and, secondly, No. 7 Squadron RNAS sent a 'bomber destroyer' to bomb the Gotha's home bases and then patrol the return route to engage the bombers. This Handley Page was equipped with extra guns and gunners and, for its time, fairly bristled with weapons. The crew saw three bombers and attacked two, damaging at least one Gotha. Despite the apparent success of the concept it was not followed up.

There was now cause for concern, and all manner of ideas were put forward as possible solutions. The basic problem appeared obvious; that of detecting the raiders and of putting a defending nightfighter close enough to make an attack (the same problem was to exercise British and German defenders alike in 1940).

The bombers were back on the last night of September and again the following night, the latter raid being the first time on which a new detection device was employed by the defenders. A 15ft-diameter spherical sound locator had been installed at Fan Bay, Dover, and could, in theory, detect aircraft some 15 miles out to sea. This was only one of a number of locator devices tried during this period, some of which appeared to show promise during trials but all of which suffered severe limitations. Many of these devices were the creation of the Munitions Inventions Department. Another technique that showed promise was to use tracker aircraft that would shadow the enemy formation and report on its position, allowing night fighters to home to the area (a technique used to great effect by the Luftwaffe in 1943–44). At the very least it showed that the Home Defence network recognised the essential role of such warning devices in any air defence network.

Meanwhile, on the Western Front, with America's entry into the war in April 1917, the hard-pressed Allies had looked forward to massive reinforcements (the first ground troops arrived in June), including elements of the United States Air Service (USAS). The only aspect relevant to this account was the eventual creation of the 1st Pursuit Group (Major Harold E. Hartney) and, in October 1917, the presence of the 185th Aero Squadron under the command of First Lieutenant Seth Low. In common with many night operating units, they adopted a bat as the unit badge. The 185th was designated as a Night Chasse (nightfighter) squadron, and was the first such unit in the USAS; its orders were to 'establish a barrage over our lines of searchlights against enemy night bombers'. The unit historian enlarged on its role:

> Night chasse work in aeronautics is only in its infancy, and as we were a new type of squadron equipped with planes that had almost gone out of service, we were confronted with numerous difficulties. In the first place our pilots

> were not trained for night flying, many of them had no experience with wing flares, parachute flares, instrument lights etc. Also, there was not enough searchlights and markers for the guidance of our pilots who frequently became lost, ran out of gasoline – and then had to make forced-landings, invariably crashing their planes.

These were all problems that the RFC had discovered in its early days of establishing a night defence in England. According to the Manual for Employment of the Air Service:

> The pilots must be specially trained in navigation by night, although this can be learned in a very short time by one of ordinary intelligence, and can be helped greatly by increasing the number of light houses, mortar signals, etc. The night fighter pilot must be imbued with the spirit of determination. He should be steady, sober, keen and industrious and so fond of flying that he seizes every opportunity to get in the air – day and night.

Such was the theory, but in practice it was a question of needs must and 'on-the-job training' for the squadron crews.

The unit's initial equipment was Sopwith Camels, and it flew its first mission on the night of 18/19 October. It was considered that a reasonable pilot should be able to see an enemy aircraft at 500–600m on moonlight nights, and at a far greater distance if searchlights were present. Patrol heights were in the region of 6–12,000ft, and normal endurance was up to 1¼hr. By 11 November the unit had flown 31 missions on 8 nights, and had also undertaken two night bombing raids. However, the squadron had one fatal crash and three wrecked aircraft. The only combat occurred on 24/25 October, when an enemy bomber was attacked five times – without result. They were proving both hard to find and hard to knock down.

Meanwhile, the German bombers were waiting for suitable moon phases before resuming their attack on England, and a major airship raid was planned for 19/20 October. Eleven airships were tasked against targets in northern England but, almost inevitably, adverse weather ruined the plan and scattered the attackers. Very little damage was caused, and although the defenders put up a record 78 sorties, no combats (other than a couple of long-range shots) took place. Nevertheless, four airships were lost, one falling to anti-aircraft fire (L44) and three others crashing in France. Another defender had been airborne with a new weapon. The Martinsyde Elephant of Captain L. Eeman, operating out of North Weald with No. 39 Squadron, had three upward-firing guns.

The bombers returned five more times in 1917, London being the target on all but one occasion. Results were poor and losses mounted, although most of these were crashes owing to poor weather. However, the first nightfighter success may have been scored by Captain G. Murlis Green on 18/19 December, flying No. 44 Squadron Camel B5192 out of Hainault. He was airborne at 1843 and 30min later was orbiting at 10,000ft over the Goodmayes area of Essex. Having turned towards searchlight activity, he then picked up the exhaust glare of a Gotha and steadily closed until he was some 30yd beneath his enemy. This particular Camel had a pair of upward-firing Lewis guns, and Murlis Green now opened fire, only to find that one gun had frozen. At that point he was forced to bank away as the Gotha released its bomb load. However, it was still held in the glare of the searchlights and fairly easy to follow. Two more attacks produced no result, but a final attack caused the Gotha to dive and the Camel to go out of control, caught in its slipstream. Murlis Green recovered from the spin but had lost sight of the bomber. After the Gotha had flown out to sea its damaged engine gave up, and the crew had no choice but to ditch near Folkestone. Credit was duly given to the Camel pilot.

The other significant event of the night was the first operational use of the Royal Aircraft Factory S.E.5a by No. 61 Squadron from Rochford, Kent. At long last, and often in the face of official disinterest, reasonable aircraft were finding their way into the hands of the HD squadrons. Another type to enter service in the nightfighter role was the superb Bristol F.2b Fighter, which was developed for day fighter use over the Western Front and went into action in early 1917. However, the continued problems that the HD units were having led to calls for the 'Brisfit' to be employed in this role. No. 39 Squadron, having struggled (and complained) for so long, was one of the units chosen to re-equip with the type. Continuing its penchant for 'home-grown' modifications, the squadron gave at least one F.2b a twin gun mount front and rear, realising the importance of having sufficient firepower to gain a quick victory, especially in the face of well-armed bombers.

The standard operating procedure in early 1918 was for no daylight standing patrols, as aircraft would be sent up as required. However, at night there were standing patrols along predetermined patrol lines. When called out, each flight sent up aircraft to fly along the designated patrol line, one of the three aircraft flying at the specified height e.g. 10,000ft, one about 500ft above and one 500ft below. It was reasoned that if another aircraft was met at that height, it must be an enemy. This somewhat dubious system was supplemented by a code of air-to-air signals, but, even so, mistakes were still frequent, as on the occasion when two squadron aircraft met over North Weald and, in the words of the report, 'had a spirited encounter', fortunately without causing damage to either aircraft.

The year 1918 opened with a continuation of the strategic bombing of London. Nineteen bombers attacked London on 28/29 January, causing £187,350-worth of damage and 67 deaths, over half of the latter being at the Odhams Press Building when this was struck by at least one 660lb bomb. The defenders put up 103 aircraft (97 from the RFC) and had a mixed night. One Gotha was shot down in flames, but a No. 39 Squadron Bristol Fighter attacked a Giant and was promptly shot down when a single burst from the bomber struck its petrol tank. The gunner was wounded and the aircraft's engine seized, but Lieutenant Goodyear was able to make a forced-landing at North Weald. It was definitely not a one-sided affair; the bombers could fight back.

The Home Defence organisation was now well established, and on nights when warning orders were issued the flying effort was impressive, if not always very effective. The basic problem remained the same; finding the enemy in the dark sky. To the ill-informed it seemed ridiculous that, with 60 or 70 fighters operating, no-one could find the bombers. In fact, the number of sightings, and combats, was increasing, and on most of the remaining raids this was to be the case.

On the night of 29/30 January Captain Arthur Dennis in a No. 37 Squadron B.E.12b caught Giant R39, one of four such aircraft attacking London that night. He attacked and scored hits on the bomber, whose rear guns in turn damaged the fighter. Nevertheless, the bomber carried on to its target, at which point it was spotted by Second Lieutenant R. Hall of No. 44 Squadron, who carried out an unsuccessful pursuit. Some minutes later, as the bomber was leaving the area, it was seen and attacked by Captain Luxmoore of No. 78 Squadron, who made a diving attack in his Camel. Yet another attack was made by a No. 44 Squadron aircraft (Captain G. Hackwill), 600 rounds being fired at longish range with no effect. The final sighting of R39 took place near the coast, when an F.K.8 of No. 50 Squadron turned towards the bomber but lost it again before reaching a firing position. However, the point to be made is not so much the lack of success, but rather the number of encounters; the defences were becoming sharper each month. Also, the bombers were forced to abandon their bombing raids. Thus, despite there being no victories, it was a definite success for the defenders.

Two weeks later, on the night of 16/17 February, a small force of Giants was over England, one of which dropped the first 1,000kg (2,200lb) bomb.

The size of bombs had been increasing over the period of the offensive, but up to this point the largest had been the 660lb bomb. The destructive potential of these larger weapons made it even more important for the defenders to try and catch the raiders before they reached their targets. A single Gotha bombed St Pancras station the following night, and none of the 69 RFC aircraft even saw the raider. There was another problem for the defenders. For some time they had to contend with being shot at by their own anti-aircraft guns (ever the lot of airmen), but now there were increasing instances of aircraft firing at each other. Added to this was the ever-present danger of mid-air collision as the fighters searched the sky for their targets, attracted to an area either by the flash of bombs of a glimpse of an aircraft in a searchlight beam. During the raid of 7/8 March, two aircraft collided (B.E.12 C3208 of No. 37 Squadron and S.E.5a B679 of No. 61 Squadron), killing both pilots.

The only other attacks in March were three by airships, all of which proved disappointing for both sides. April was a quiet month over England, the bombers being employed against targets on the Western Front. The night of 19 May brought the final aeroplane attack on London. It was the largest yet, with 38 Gothas and 3 Giants taking part, although ten of the Gothas turned back early. The Royal Air Force (which had been formed on 1 April), put up 88 sorties and had a productive night. RAF patrols were airborne from around 2253, and the first bombs fell on London at 2330. Just a few minutes before that, the first bomber fell victim to Major Brand, the CO of No. 112 Squadron, flying Camel D6423. Brand was on a patrol line between Throwley and Warden Point on the Isle of Sheppey when he saw a Gotha over Faversham, picking out the aircraft by its engine exhausts. He came under attack from the Gotha's front gun and returned fire, damaging its starboard engine. As the bomber attempted to evade, Brand got on its tail and put in a concentrated burst of fire, causing it to catch fire and fall apart.

Another was brought down over East Ham by a Bristol Fighter of No 39 Squadron flown by Lieutenant A. J. Arkell, with Air Mechanic T. C. Stagg as gunner. Arkell closed with the formation of bombers at 10,000ft over London and vigorously attacked one of the Gothas. The aircraft broke formation and tried to escape by diving down towards the city, hotly followed by Arkell. The pursuit continued until the aircraft were only 1,500ft above London, when at last the Gotha was hit again and again and, with her pilot either dead or wounded, crashed into the ground. A third went down shortly afterwards. This aircraft may already have been damaged by another fighter before eventually falling to Lieutenants Edward Turner and Henry Barwise in a No. 141 Squadron Bristol Fighter. A fourth Gotha was lost when, owing to an error by the crew, it crashed at St Osyth. It had been a terrible night for the raiders, even though they had caused a fair degree of damage. The night defences were now too strong and the bombers never returned, all of their final endeavours being on the Continent.

By April, after an impressive start due to massive reinforcement from the defunct Eastern Front, the German spring offensive had forced salients in the Arras and Ypres areas. The Allied front then stabilised, partly owing to an influx of American ground forces and the use of air power. In response to the Allied build-up, the German heavy bombers increased their night attacks against targets behind the battle front, in what was a total misuse of strategic air power. A few raids were made against cities, such as the attack on Paris by eleven bombers on 1/2 June, but most of the effort was expended against targets within 50 miles of the front lines. This is turn led to an increase in the nightfighter defences, which up to that point had largely comprised coastal units whose primary role was protection of port areas and countering the raiders bound for England. Most defence in the frontal areas relied on day fighter units that, from time to time, tried their hand at night work.

A detachment of Camels from No. 58 Squadron moved to Fauquembergues on the night 'route' to Abbeville to intercept the bombers. At 2350 on 31 May 1918 Lieutenant C. Banks saw a bomber in a searchlight, closed to 25 yards and opened fire, sending the enemy machine down in flames. This was one of the few successes, and it was realised that greater efforts needed to be made. The very experienced nightfighter pilot Major Murlis Green was instructed to form a specialist squadron for France. This unit, No. 151 Squadron, moved to its new base in mid June and was soon in action. In its first five months of operation the squadron shot down 26 enemy aircraft for no loss (except the usual round of night-flying accidents). September saw particularly heavy air action, day and night. In the latter part of the month No. 151 Squadron accounted for at least six of the bombers, three of these falling on the night of the 21st. Included in the Squadron's total were two of the R-planes; R.43 falling to Captain A. Yuille near Talmas on 10/11 August, and R.31 going down near Beugny on 15/16 September: 'Lt S. Broome saw a giant enemy machine held in our searchlights which he attacked, firing 500 rounds altogether. The enemy machine burst into flames and fell on our side of the lines.'

While the German night attacks were significant, it was the Allies who were establishing the future of such action with the creation of the Independent Air Force. With its five squadrons of Handley Page O/400 night bombers (plus Airco D.H.4 and D.H.9 day bombers), the IAF was created to attack such targets as troop concentrations and, especially, lines of communication. Mobility of forces was vital, and for the Germans this largely centred on the excellent rail network, which thus became a primary target. In addition to the night efforts of the IAF, there were those of the French Night Bombardment Group, equipped with Caproni biplanes. In the face of this night effort the German defences were ill-equipped, reliance being placed on anti-aircraft guns and searchlights.

Whilst it is true to say that the German nightfighter 'force' did not come into existence until 1918, there had been night combats, and successes, the previous year: the first real night air-to-air victory had been scored by Leutnant Frankl of Jasta 4 when he shot down a B.E.2b of No. 100 Squadron on 6 April 1917, followed up with a Nieuport the next night. The first success against the Allied bombers attacking Germany itself went to Kampfeinsitzerstaffel 2 (Home Defence Squadron 2) on 8 August 1917, the bomber falling near Saargemünd. However, a recognisable system was not created until May 1918 with the co-ordination of aircraft, searchlights, guns and an air raid reporting service. Amongst the first units to put the system into operation was Jasta 24, which was deployed to protect German ground forces on the Western Front from night bombing. Its Albatros D.Va fighters flew the first such operation on the night of 21/22 May – though without success. However, the next night the squadron commander, Leiutnant Thiede, shot down three bombers. In similar vein, Jasta 73 established a nightfighter reputation, scoring well in the August–September period, including a score of five French Voisin Vs falling in a single night to Leutnant Gerhard Anders.

Meanwhile, the final bombing attack on England was a parting gesture by the airships on 5/6 August. Only five airships took part and, although their targets were in the Midlands, none crossed the English coast. Once again bad weather played a part in spoiling the attack. L70, the very latest Zeppelin in the German inventory, was shot down into the sea off Wells-next-the-Sea, the victory being credited to Major E. Cadbury and Captain R. Leckie in a D.H.4 from Yarmouth, though it was also claimed by another crew.

The strategic bombing offensive against England was over. Lessons had been learned, although many of those would be forgotten in the years following the 'war to end all wars'. The art of the nightfighter had developed as a result of a series of airship and aeroplane raids against England; raids that held little military significance but carried huge political (and psychological) gain.

Chapter 2

THE INTERWAR PERIOD

The end of the First World War brought a rapid disarmament programme among the western nations, with little thought being given to the future, or so it seemed. There was, understandably, a war-weariness and aversion to things military, plus an urgent economic need to return to a peace-based economy. The nascent air arms were among the most seriously affected by this rush to disarm. This was partly due to internal military wranglings; very much so in the case of the RAF, its older (and more influential) Army and Navy brethren seeing only a limited need for air power, and decreeing that such small amounts as were necessary would be better integrated within their spheres of control, rather than operating as a separate entity. There were powerful arguments on all sides, and in the wranglings many of the lessons that should have been absorbed were either glossed over or ignored. However, a number of studies were undertaken to assess the contribution made by aircraft in this, their first conflict. All agreed that reconnaissance was vital to military operations. Most were convinced that the bomber had played a role, if only in influencing public and political opinion. It was to be this latter strand that was to dominate air power doctrine for the next two decades.

This is not the place for a discussion on the development of air power theory in the 1920s and 1930s; the commentary must be restricted to a few generalised statements that have a bearing upon the study of the nightfighter. As was made obvious during the analysis of the birth of the nightfighter, the role evolved as a direct result of the need to counter an airborne threat; in the first instance, the strategic bombing of England by the German airships. Thus, in theory, if there is no threat, there is no need of a counter. This may sound obvious, but such was the level of discussion taking place during this period. In the early 1920s the air power theorists decreed that the bomber would be the war-winning weapon. It was able to fly over the enemy's ground forces, no matter how powerful they were, and strike at the heart of his government and economy. Here was the military tool that would make all others redundant. What country would risk war if its cities and seats of power could be laid waste almost at whim?

These theorists did not envisage the small-scale efforts that the Germans had employed against targets such as London but, rather, massed air fleets capable of destroying a city. It was powerful stuff, and it found a ready audience. The only way to counter the bomber was with a fighter defence, but, the theorists argued, it would be impossible to shoot down even a small fraction of the bombers. More than enough would get through to cause wholesale destruction. First there was the problem of finding the bomber, then of shooting it down in the face of its considerable armament. If one accepts the argument that the bomber will get through to its target regardless of the defending fighters, there is no need to make life difficult by attacking at night. Far better to attack in daylight, when you can be sure of finding and hitting the target.

With this as the basic doctrine of all the major air arms, there was no real requirement for either a night bombing capability or a nightfighter capability. However, a number of specific designs were put forward during this period. One was the American Curtiss PN-1 (PN for Pursuit Night), a single-seat biplane that leaned heavily on German First World War designs. One example was built and then abandoned.

Nevertheless, the writing was on the wall; the night skies could be used. June 1922 saw the first commercial night flight between European capitals, when a Farman Goliath of Grand Express Aériens flew from Le Bourget, Paris, to Croydon, London's new airport. Anything that a civil aircraft could do could be repeated by a bomber. Had the lessons of the Gotha and R-plane raids been so soon forgotten? The following year, the United States Mail Service flew trial night operations using landing grounds marked by searchlight beacons. The night hours were gradually being opened up. A few years later, in July 1926, Blackburn Dart N9804 landed at night on HMS *Furious*, the first such night landing on an aircraft carrier. However, those who considered the night bomber threat even a remote possibility were of the opinion that the day fighters would be able to take on this role if required. The lessons of the First World War had certainly not been learned in this respect, and the oversight was not to be corrected until 1940.

The RAF acquired some superb fighter aircraft during the 1920s, a number of them being given a notional night-flying capability. They included the Hawker Woodcock (which has been claimed to be the first aircraft designed as a nightfighter) and the Bristol Bulldog. The Woodcock was put forward for Specification F.25/22, the first direct requirement to call for a nightfighter. Initial problems resulted in the Woodcock II, which made its first flight (J6988) in August 1923. The first unit to fully re-equip with the type was No. 3 Squadron at Upavon in May 1925, but even though the squadron was designated a night unit, pilots spent a great deal of time on daytime operations. The aircraft, with its maximum speed of 143mph and excellent handling characteristics, was well liked by the pilots.

In due course the Woodcock was replaced as the RAF's 'nightfighter' by the Gloster Gamecock, although this was very much primarily a day fighter and did not meet the same specific requirement. It entered service with No. 43 Squadron in March 1926, and in terms of overall effectiveness, with its twin Vickers guns and a top speed of 155mph, was no great improvement over the fighters in use at the end of the First World War. One of the many aspects of night warfare that few had yet addressed was the need for heavy armament to ensure the enemy's destruction in the brief engagement.

Air Ministry Specification 9/26 called for a single-seat nightfighter, although this was later changed to day fighter/nightfighter, to replace the Woodcock. A number of aircraft were proposed, but none went into production and no progress was made until the issuing of Specification F.7/30, to which a number of companies responded in 1931. Timescales called for the best designs to go to prototype stage for a competition in 1934. The specific nightfighter requirements included a steep rate of climb for night interception and the provision of full night-flying equipment. No mention was made of general performance or weight of armament; points that most nightfighter pilots of the First World War would have stressed. Twelve designs were submitted, including the Gloster Gladiator.

By the mid-1930s the Air Ministry was issuing Specifications covering a wide range of aircraft types, many of which included nightfighter potential. Specification F.37/35 called for a day/night fighter armed with four 20mm cannon (quite a change from the requirements of just a few years earlier) to give 'striking power superior to that of the eight-gun fighter'. Among the proposals were the Westland P.9 Whirlwind and the Bristol Type 153. British aircraft manufacturers responded magnificently to the rearmament programme of the late 1930s.

Developments were also taking place across the Channel. Under the terms of the Treaty of Versailles, Germany was not permitted an air force, but planning for such an air arm was well under way in the mid 1920s, the booming spheres of civil and private aviation providing excellent 'training schools'.

In 1926 the German military identified requirements for four primary types, one of which was the NAKUJA (Nachtjagd und Erkundungs-flugzeug) nightfighter. All trials took place in secret at the Lipetsk base in Russia, and in the 1928 test and evaluation programme the BFW M-22 was put forward for the NAKUJA requirement. This proved generally disappointing, as did the Junkers K-47 in the 1930 test, but they showed that the military was at least addressing the problem.

The RAF's most advanced bomber, the Fairey Night Bomber (later named Hendon) first flew in November 1930, having been originally designed to meet a 1927 specification for a heavy bomber. Entry to service was with No. 38 Squadron in late 1936, by which time the bomber was dangerously out of date. Its relevance to this study is its definition as a night bomber, the implication being that night warfare was a possibility once more. It was woefully antiquated, with a top speed of only 155mph and a ceiling of 21,500ft. Its defensive armament comprised three Lewis guns.

As far as its strategic bomber force was concerned, the RAF put its faith in aircraft such as the Vickers Wellington, faster, more heavily armed and due to enter service in the late 1930s. With this type the tactical concept remained unchanged; go by day in a tight defensive formation.

The various RAF Expansion Plans of the early 1930s continued to place greatest reliance on the bomber as the main element of its strength. However, Britain and France received a distinct shock in 1935 with Hitler's formal declaration of the existence of the Luftwaffe and his announcement that its strength was equal to those of the other nations. Great delight was taken in parading the latest aircraft, such as the Heinkel He 111 and Dornier Do 17 monoplane bombers, both advanced designs with reasonable performance, considering that the RAF's primary bomber was still the lumbering Handley Page Heyford biplane, although other types were under development.

As aircraft performance increased, so, too, did airfield requirements, especially if night operations were to be taken into consideration. Airfield development was an integral part of the air forces expansion programmes; many airfields were, at last, being provided with night-flying equipment.

One of the inherent problems during the First World War had been the lack of an effective detection and reporting organisation, despite the adoption of sound locators in the latter months of the conflict. June 1922 saw the creation in Britain on the Observer Corps, out of the Defence Corps, although this organisation relied on visual acquisition, recognition and tracking. While this proved invaluable during 1940 for the daytime battle, it was distinctly limited in its night application. The only 'technology' at this period was that operated by the Army Acoustic Section, essentially the same type of concrete 'mirrors' as used previously. While these devices were reasonable in providing azimuth of a target, they gave no indication of height and were therefore of limited use at night for interception data. However, experiments examining the properties of radio waves were already being carried out by the newly-created Radio Research Board (part of the Department of Scientific and Industrial Research), headed by Admiral Sir Henry Jackson. This basic work was to provide essential background data for the development of Radio Direction Finding (RDF) in the 1930s.

The British Air Defence Plan was based upon that employed during the First World War. It comprised an Aircraft Fighting Zone 15 miles wide and 150 miles long, from Duxford and around London to Devizes in Wiltshire. This was divided into ten sections, each 15 miles wide and with one or more dedicated day fighter squadrons, with associated searchlights and anti-aircraft guns. There was an additional ring of searchlights and guns around London. The Aircraft Zone was positioned 35 miles from the coast, a distance based upon the time it would take the fighter to climb to 14,000ft. Initial detection of raiders depended upon visual sighting by the Observer Corps, plus a limited number of 'acoustic mirror devices'. The only real refinement of this system over that of 1917 was the co-ordination provided by the sector operations rooms. Each HQ received information from the observation units and displayed raids on a plotting table, thus allowing the overall picture of the air situation to be seen by the controller. However, it was not until new radios, such as the TR.9, were introduced in 1932 that reasonable ground-to-air radio communication, as opposed to W/T, was possible. Even then the effective range was only 35 miles. The weak link remained that of detection. If the enemy could not be accurately located, it was almost impossible to effect an interception. This was borne out in the July 1934 annual air exercise, when at least half of the day bomber formations reached their targets without being intercepted by fighters. If the fighters could not find bomber formations in good weather by day, what chance would they have at night? Not that this was of any concern at the time.

At the first meeting of the Committee for Scientific Survey of Air Defence, in January 1935, the problems highlighted by the previous year's exercise were discussed, as were the prospects of any scientific breakthrough that might provide an answer. The Committee members consulted Robert Watson-Watt, head of the radio research branch of the National Physical Laboratory, as to the feasibility of using radio waves. A month later he presented his thoughts on how to use such radio waves to detect aircraft, the principle being to 'bounce' the waves off the aircraft and pick up the echo. Within a matter of weeks an experiment had been arranged using the BBC's transmitter at Daventry. The idea was for an aircraft to fly through the centre of the transmitting beam while Watson-Watt and his colleagues attempted to detect its presence on a cathode-ray oscillograph. It worked as planned, the passage of the aircraft causing a blip on the equipment. All that was needed now was high-level support for development of the technique, and that came from Air Marshal Hugh Dowding, the Air Member for Research and Development. It is unlikely that the concept, which was only one of many scientific devices seeking backing, would have progressed as rapidly as it did without his support.

The radio research station at Orfordness became the experimental site for the development of RDF, and throughout the summer a number of trials were conducted. Tracking ranges of 40 miles were soon being achieved, and it was obvious that here was the solution to the problem; all that was needed was time to develop and introduce the equipment, and integrate it into the Fighter Command system. In September 1935 the Air Defence Sub-Committee acquired Treasury funding for a chain of RDF stations along the east coast; a remarkable achievement so early in the development of the technique. The next three years saw a number of technical developments of the equipment, many individuals making invaluable contributions to the work, so that by mid-1938 the completed stations were functioning reasonably well. The Bawdsey site became the focus of a great deal of German intelligence gathering by land, sea and air, the latter involving flights by the *Graf Zeppelin*, successor to the airships that had bombed England twenty years before.

When the Luftwaffe visit to the RAF took place in October 1937, the centrepiece of which was a display at Mildenhall, it was revealed by General Milch that the Germans considered themselves to be ahead in the development of the use of radio waves for detection. This was not a complete shock to the British, as reports had filtered through about such work. It seems that the Germans had a working RDF system in 1934, developed by the Navy Signals Research Unit. However, subsequent development was slow and the British work soon forged ahead.

The term RDF was retained until 1941, when the system became known as radiolocation, an accurate description of the process. However, to accord with American nomenclature it became radio direction and ranging (radar) in 1943.

Meanwhile, as part of the German re-armament programme, an aircraft requirement was issued in October 1932 which, among other items, called for 'Rearmament aircraft II', of which one variant was to be a two-seat day/night fighter. The requirement was modified the following year to favour a 'Flugzeugzerstörer' (aircraft destroyer), with cannon armament and 'measures to reduce noise and prevent flames from the exhaust so that the aircraft will suit a nightfighter application'. This was an aspect of night operations that always caused major problems; many combat reports from both world wars state that the enemy's presence was initially betrayed by the glow of the engine exhausts. The fly-off competition in the spring of 1936 had three entrants, but by the following year it had been decided to proceed only with the Messerschmitt Bf 110, initially in day fighter guise. Events were to show that the aircraft was best suited to its 'secondary role'. The same year saw the first German nightfighter trial, at Berlin in May when aircraft of II/JG132 co-operated with searchlight units to counter 'attacking' bombers. Further trials took place the following year and confirmed that the basic procedure of 'illuminated nightfighting' was practicable. The subsequent Air War Academy report suggested that nightfighting was practicable not only in home defence but also when used by long-range fighters against the enemy's home bases. Despite these favourable trials, however, very little was actually done: the Luftwaffe's offensive doctrine tended to ignore defensive concepts.

The outbreak of the Spanish Civil War in July 1936 was to bring the first application of the new air power doctrines and provide a 'testing ground' for aircraft and weapons. Both tactical bombing (close air support) and strategic bombing were employed, especially by the Luftwaffe's 'Condor Legion'. While most air action took place by day, there were a number of night operations. Among the Russian aircraft deployed to Spain were fighter units tasked with the air defence of Madrid, the capital coming under increasing pressure in the first half of 1937 as the Nationalist forces continued their advance. After the shock bombing of Guernica on 27 April there was a great fear that Madrid would receive similar treatment. However, as far as the night war was concerned, the first victory was scored on the night of 27 July, when Mikhail Yakushin, flying (probably) an I-16, shot down a Junkers Ju 52. The following night one of his colleagues, Anatoli Serov, brought down another. Both pilots were awarded the Order of the Red Banner for 'sustaining the morale of the beleaguered capital'.

The scale of night operations during the Spanish Civil War was such that it had little impact on the doctrine of the participants or upon the many other air forces that were observing this first 'modern' air war. Many other more important lessons were being learned.

Since the initial proving of the RDF principle in 1936, progress had been fairly rapid, although many problems still remained. By mid-1937 three stations were in operation, at Bawdsey, Canewdon and Dover, along with an experimental filter station at Bawdsey. The development of the last of these added a new dimension to the system by providing, as the term implied, a filtering of the mass of information from the various sources so that the controllers could be provided with a simpler, more accurate air picture on which to base their operations. The earlier problem of track discrimination had virtually been solved, but height prediction remained a significant problem. The system's first major test came in the 1938 Home Defence exercises, and in general terms it appeared to work well, some 75 per cent of attempted interceptions (day and night) proving successful. New RDF stations (codenamed Chain Home, CH, and Chain Home Low, CHL) were constructed in a plan to create unbroken coverage around the east and southern coasts of Britain. At the same time an extensive programme for the construction of Command, Group and Sector operations rooms was under way.

The enemy could now be located by day or night; radar made no real distinction between the two. Defending fighters could be positioned so that by day they could acquire the bombers visually, but what about the night situation? Would the fighter be able to pick up its targets? Among his early proposals, Watson-Watt had included thoughts on an airborne version of RDF to prevent reliance on searchlights or good visibility. However, because priority was given to the ground stations, little work was carried out on developing equipment small enough to be carried by aircraft. June 1937 saw an experiment whereby the 'fighter' had a receiver that could pick up the transmissions (and echoes) of the CH stations. The principle appeared to work, but it was not a satisfactory solution and was abandoned. Not until late that year was Airborne Intercept (AI) equipment developed, using the technique of lobe switching to determine the azimuth of the target. Although this worked, there were still problems to be overcome, mostly concerning the receiver part of the system, and it was mid-1939 before a successful airborne installation was under trial.

Meanwhile, an Air Fighting Committee paper of October 1938 reported on a French system of night interception without the aid of searchlights ('Chasse obscure'): 'A fighter pilot can usually see an unilluminated bomber at night providing he is below it; the range varies between 600 yards and 6,000 yards depending on conditions of visibility.' The basic French system was intended to comprise a sector 15km wide by 20km deep, at the front edge of which was a 'sensing device' which would be triggered by the passage overhead of an aircraft. Some 5km further back would be a series of sound locators whose 'trained personnel would be able to predict the enemy's height to within 100m'. The pilot of the nightfighter would then be given a height at which to fly so that he could take advantage of the improved detection by being below his target. Another 5km back would be a further batch of sound locators to work out the track and speed of the enemy, this information being passed through to a control centre that would also be responsible for positioning the nightfighter using direction finding (D/F).

The reliance on sound locators was widespread, and was to survive as a basic option until the general introduction of RDF, which was to prove the real key to the problem of target location and fighter interception. French air strength looked impressive on paper, but the French, like the British, had been slow to adapt to changing doctrine and capabilities.

In mid-November the RAF's Air Fighting Committee issued Report 57, discussing nightfighter operational techniques:

> A few recent experiments carried out at night have indicated that it should be possible to navigate fighters by means of D/F intercept techniques to within about 4 miles of an enemy – providing sufficient information regarding track and height of the enemy is available. The accuracy of interception by D/F at night will be such that fighters will usually be unable to sight the target unless the latter is illuminated by some means or unless further detection aids are provided.

The report went on to look at the options for such detection aids:

> Use of AI – it is suggested that at least one aircraft in the fighter formation should have AI equipment in order to determine whether this apparatus would enable the aircraft to make visual contact with a target which is not illuminated.
>
> Searchlights – on a clear night, the intersection of searchlight beams should be sufficient indication of the position of the target. Experiments should be made to determine, under various conditions of visibility and at various altitudes, the range at which the intersection of searchlight beams is visible and the range at which a target illuminated by searchlight beams is visible.

> Owing to the limited number of fighter units which can be operated by D/F techniques, it would seem desirable for fighters to fly in formation at night; therefore, aircraft will need to be equipped with station-keeping lights and IFF [identification friend or foe].

The trials did not take place until the following spring and summer. The first series were flown between 12 April and 4 May, with a No. 38 Squadron Wellington as the target and a No. 111 Squadron Hurricane as the fighter, the bomber being finished in the latest night camouflage scheme of matt black. After a number of sorties the general conclusion was that:

> . . . the range at which pilots can see an unlit bomber whose position has been indicated, depends upon the experience of the pilot concerned. The upper limit appears to be about 600 yards astern and below, and 3,000ft directly below. Aircraft engine exhaust can be seen up to one mile away. The average range for detecting searchlight intersection is 15–25 miles and 8–15 miles to distinguish an aircraft being held by searchlights.

So far, so good. The second phase was conducted between 8 May and 7 July, the same aircraft being involved:

> It is likely even under good conditions that searchlights will only help the fighter by 'flick-overs', to close with and attack a bomber in the dark. If, however, the lights continue to hold the bomber or make a series of flick-overs without dropping behind it while the fighter closes until he can see the bomber itself, it is practically certain that the fighter will be able to deliver an attack. The approach to and attack on both a lit and unlit target should be made from behind and below, fire probably being opened at a maximum of 200 yards. Fighter patrols should be placed at least 10 miles behind the front line of searchlights in order to avoid confusion and to allow the fighter to go forward to intercept when a pick-up has been made. Use of the aircraft landing light as a searchlight did not prove successful.

The final comment is interesting, and demonstrates that the possibilities of airborne illumination were already being considered. The subject was addressed by the RAF somewhat later with the introduction of the Turbinlite system. During this series of trials 143 attempted intercepts were set up, of which 43 resulted in visual pick-ups and 'combats'.

So, by the late 1930s there was a realisation that nightfighters would have a role to play, and that in the absence of such technology as airborne RDF, reliance would have to be placed on fighter pilots acquiring their targets using the 'mark one eyeball', albeit with whatever assistance ground units could provide. Searchlights appeared to hold the most promise, and a combination of fighter and searchlight was to remain an important tactical concept for night combat throughout the coming war, with ground radar giving the fighter a general target position.

By the spring of 1939 the RAF's system of RDF stations was expanding rapidly, although there were still a great many gaps – only twelve of the planned nineteen stations were operational. The conclusion after the final peacetime Home Defence exercise, in August 1939, was that 'the system, although doubtless capable of improvement as the result of experience might now be said to have settled down to an acceptable standard.' Fortunately it was to be given another year in which to expand, develop and be refined before being put to the ultimate test by day and night.

The next breakthrough came in May, with the installation of an experimental AI set in a Fairey Battle. This aircraft carried a transmitting aerial in the port wing and four receiving aerials (for discrimination), two for azimuth and two for elevation. The AI operator was provided with two cathode ray tubes (CRTs), one for the azimuth readout and one for the elevation. Trials using a Handley Page Harrow as a target aircraft proved that the system was usable, if a little unwieldy and prone to equipment failure. Suggestions for a combined CRT for use in single-seat fighters were put on hold when the C-in-C Fighter Command, the post to which Dowding had moved, supported the use of specialist two-seat nightfighters.

The time for theory was rapidly passing. Now came the acid test of a major conflict.

Chapter 3

THE FIRST TEST

'The German Air Force very quickly reacted to our night bombing offensive. As early as September 1940 they had developed a nightfighter force of 120 aircraft out of their day fighter units. By the end of 1941 the effectiveness of the enemy nightfighter system resulted in such heavy losses to our bombers that it was decided to restrict bomber operations over Germany until the spring of 1942. By June 1942, the number of nightfighters had risen to 250, by July 1943 it was 550 and by the spring of 1944 over 800. The nightfighter force remained efficient to the very end of the war. In the face of these defences losses were considerable. The rate of night bomber losses was highest in 1942, with a loss rate of 4.1 per cent. A casualty rate of 4 per cent meant that on average there was a turn-over of aircrews in squadrons in about four months due to casualties.'

So said Air Chief Marshal Sir Norman Bottomley during the RAF's postwar conference, Exercise Thunderbolt. This study was only concerned with the progress, effectiveness and problems of the Allied Strategic Bomber Offensive, and so only paid attention to German developments in night defence. The nightfighter story of the Second World War has two distinct parts, the British and the German. Although the American, Russian and Japanese air arms developed nightfighter aircraft and tactics, these were never on the same scale, complexity or effectiveness as those in western Europe. In this major part of the story of nightfighting it is only right, therefore, that most attention is focused on those key developments. It is a complex pattern, although each nation tended to go through the same learning process and, in general terms, arrive at similar solutions.

It is a frequently-held opinion among the uninitiated that it should be easy to spot another aircraft in the night sky. However, any aviator will tell you this is far from the truth. It is often hard enough to spot another aircraft on a bright sunny day, even when the ground radar unit is passing the relative position of the other aircraft. At night aircraft display high-visibility strobe lights, and these flashing red lights are excellent at revealing the aircraft's position, but switch them off (as one does in a hostile situation) and the aircraft vanishes into the pitch black night. Even if you know where it should be, having been watching its lights, it has now gone. The situation does, of course, depend upon the amount of ambient light (moonlight or artificial) and the reflectivity of the target, plus the skill and experience of the seeker. Air Fighting Committee Paper No. 112, dated 28 May 1941, expressed the following opinion under the heading 'Notes on night visibility':

> . . . a study of the night search and the factors which aid successful night search are of vital importance for the night fighter. The night fighter pilot's task calls for precision and care for detail both in preparation for a sortie and when engaged in search for the enemy. Be night adapted – don't let a dirty windscreen spoil your efforts, look to the detail – study the weather and think out how you intend to apply it to your tactics before each sortie. Remember, once you have picked up your hun it is easier to hold him in view than to start your search afresh – therefore, hold him! A moving target is much easier to see than a still one; make relative movement, therefore, by gently rocking your aircraft when searching. Don't search vaguely into space. Try to visualise your target at the range at which you expect to find it and search accordingly.

Night adaptation is a very important aspect, and is more accurately expressed as night vision. The human eye and brain are a remarkable combination, and can adjust to a great many situations, including seeing in the dark. Night vision develops over a period of time. You can test this by standing in a dark room and noting, over a period of 20–30min, how much additional detail becomes visible. If someone then comes into the room and switches on the light, your night vision is gone in a fraction of a second, and it will take another 20–30min to re-establish. This is a simplistic analysis, and individuals vary both in the speed and perception of their night vision ability. Nevertheless, it is a valid notion, and one that should be borne in mind when we examine, later in this study, certain tactical failures such as the adoption of airborne searchlights with the Turbinlite Havocs. Even with an established night vision, the range at which aircraft can be observed is still minimal, assuming that they are not showing any lights or reflecting any other form of light. The most frequent 'give-aways', as will be emphasised in many of the subsequent accounts of combats, were either the switching on navigation lights when returning to home base or the glare (or even minor glimmer) of engine exhausts.

At the outbreak of war the developments by Watson-Watt and the Bawdsey Research Station in respect of airborne radar were still very much at the trials stage, although the concept had certainly been proved. The night defence of Britain still relied on the 'old firm' of searchlights, anti-aircraft guns and visual ('cat's-eye') fighters, the latter being single-seat daytime aircraft. However, the first steps had been taken to bring the new technology into play. As AOC Fighter Command, Hugh Dowding had, for a variety of reasons, supported the employment of twin-engined aircraft with the new airborne RDF (soon to be termed AI) equipment, and with a specially-trained operator, for nightfighter duties. Others had favoured the development of a smaller set that could be fitted into single-seat fighters. Priority was given to equipping the Bristol Blenheims of No. 25 Squadron, this unit having received its first Blenheim IFs in December 1938.

There were still huge problems to be overcome, as well as new tactics and techniques to be developed. The basic purpose of the AI set was to provide the bridge between the intercept position achieved by the ground controller and visual acquisition by the pilot, as the accuracy of ground radar could only put the nightfighter in the general area of his opponent. One of the strongest arguments for a multi-crew nightfighter was the advantage of more than one set of eyes scanning for the enemy.

Among the many other problems and priorities in the winter of 1940, night air defence was a minor consideration. Far more important was the need to have the capability to defeat any major daylight attack. The radar network and its associated control and reporting system was at last coming

BRISTOL BLENHEIM

The Bristol Blenheim was one of a number of 1930s 'private venture' designs that proved so valuable to the RAF when re-equipment strategies were eventually put in place in the late 1930s. With the adoption of plans for an airborne RDF system, the Blenheim became the natural choice for a platform. The first RAF unit to re-equip with the Blenheim IF nightfighter was No. 25 Squadron in December 1938, then based at Hawkinge in Kent. This variant had its normal armament augmented by an underfuselage gun pack with four Browning machine-guns. As the first AI-equipped fighters, Blenheims played a major role in developing the equipment and tactics that would provide the basis of the RAF's nightfighter defences. The first night victory was achieved on 22/23 July 1940. Although a number of Blenheim IVs were also converted to the nightfighter role, most were used on operational trials such as that conducted by No. 600 Squadron.

Blenheim I data

Powerplant: Two Bristol Mercury VIII. Crew: Three. Length: 39ft 9in. Wingspan: 56ft 4in. Maximum speed: 260 mph. Ceiling: 25,500ft. Armament: Gun pack with four Brownings, single forward Browning, single Vickers 'K' gun in dorsal turret.

together but, as yet, there was little real thought of 'fine-tuning' this for use at night. However, throughout 1940 the radar network was extended and improved, both aircrew and controllers gaining valuable experience. Problems with early sets included interference with the TR9D radio, preventing air-to-ground communications. The only way to overcome this was to leave the AI set switched off until the ground link, and data from the controller, was no longer required.

Using AI Mk. I, trials at Martlesham Heath in Suffolk continued. The first attempted night intercept of German aircraft was directed against the minelaying aircraft that were becoming an increasing menace off the East Coast. These minelayers from KG4, KG26 and KG30 were almost impossible to intercept using existing tactics. On 22 November, 11 Group stated:

> German aircraft have been active off the east coast each evening after sunset, believed to be minelaying. It is the intention to intercept and engage these aircraft to seawards employing AI Blenheim fighters operating under the control of coastal radar stations that have R/T and trained controllers.

Blenheims of No. 29 Squadron also attempted to attack the minelayers. On the night of 10/11 December two aircraft tried a new technique. Having been ordered to investigate reports of a seaplane on the water near Friston, the aircraft left Debden, Essex, with the intention that one would run over the area at 4,000ft to drop a flare, in the light of which the other would investigate and, if appropriate, carry out an attack. The weather was poor, and although Sergeant Bloor dropped his flares, Flight Sergeant Packer was unable to find any trace of an aircraft. The weather continued to deteriorate and both aircraft headed home; unfortunately L6740 crashed, killing Packer, although his gunner escaped with minor injuries.

Two aircraft were detached from the Radar Flight at Martlesham to work with Bawdsey, and two from No. 25 Squadron were detached to Manston to operate with Dover and Dunkirk. This 'force' was further enhanced on the 27th by the detachment to Manston of three aircraft from No. 600 Squadron, flying Blenheim IVs (although at this time the squadron's main strength comprised Blenheim Is).

Over the next three months these detachments gained an enormous amount of experience, as did the associated radar units and controllers, but there was immense frustration all round at the lack of results. The enemy minelayers continued to operate and sinkings of coastal shipping continued to rise, leading to the adoption of standing patrols. The Air Historical Branch narrative sums up the problems: 'Neither effort nor ardour could overcome the technical weaknesses inherent in the primitive airborne radar equipments of those days, nor the drawbacks in the sets used for ground control.'

The problem with AI Mk. I was its limited range. It required too accurate a set-up from the ground-controlled interception (GCI) unit, and at this stage of the war that was not possible. The conclusion was that 'the Mark I equipment is of no operational value, but the sets are being used for training.' Nevertheless, the important first steps had been taken; although there would be problems and setbacks in the coming months, airborne radar was a proven concept that would eventually dominate the night war.

The situation in early 1940 was the same as it had been the previous year, expressed by one ex-nightfighter aircrewman thus:

> Crews launched themselves hopefully into the unfriendly darkness . . . chased around after rumours and found nothing, and then had to grope our way back through the weather to ill-equipped airfields, no navigation or landing aids. But at least we were learning about night flying.

These sentiments were the same as those expressed by the pioneer Home Defence aviators of 1916 as they struggled in the night defence of London.

The trials that had started at the end of 1939 continued, the No. 600 Squadron detachment operating with Foreness CHL site trying to determine the best method of ground control by looking at various intercept parameters and geometry. One Blenheim would act as the target, flying some way out to sea before turning back in for its 'attack', and the other would be kept on alert on the ground and scrambled to intercept. Among the major problems were those of poor R/T; not least that of the fighter pilot leaving his R/T switched to 'transmit' and thus being unable to receive control during the final stages. These Practice Intercepts (PIs) provided a mass of data that was to prove invaluable in the development of the RAF's GCI technique, which became the basis of effective night defence later in the war.

In common with the other AI-equipped Blenheim units, No. 25 Squadron was heavily involved with 'RDF trials' in the early part of 1940 in an endeavour to develop the potential of the new equipment. It was still very much a mystery to most crews, the Special Operators being viewed as 'magicians' (but often being referred to in an assortment of unflattering terms). In conversation with the author, John Cunningham, who was to go on to become one of the RAF's greatest nightfighter exponents, recalled the arrival of the AI aircraft: 'The equipment looked a bit like a "what the butler saw" machine and at first we had no real idea of what it was meant to do. It was all something of a mystery.'

While the AI aircraft had to cope as best they could with the first-generation equipment, the scientists were making rapid progress with the next generation of sets. The spur was the invention of the magnetron. In the early part of 1940 a team at Birmingham University, led by Professor Randall, had created this small but powerful device, which was to become the heart of most airborne radar. When a scientific team went to visit their colleagues in America later in the year, they took with them a magnetron, thus aiding and accelerating American developments.

March 1940 brought a number of significant events, such as the establishment of the Night Interception Committee (NIC), under the direction of Air Marshal Peirse, Deputy Chief of the Air Staff. This organisation was to be of key importance in the decision-making process over the next year or so. At this inaugural meeting the conclusion was reached that: 'defence against night attack was one of the biggest problems we had to face. Even if the enemy began by raiding in large numbers by day our ground defences would force him to adopt night bombing.' This was an accurate prediction, no doubt partly based upon the RAF's own experience of day and night offensive operations in the first few months of the war. The solution, however, was not as simple, and there was much yet to be learned.

That same month a new specialist unit was formed at Tangmere: the Night Interception Unit (NIU, soon to renamed the Fighter Interception Unit, FIU). It was equipped with six AI Blenheims, and in addition to the usual aircrew and groundcrew establishment it had a scientific and technical staff, along with laboratory and workshop facilities. The FIU provided the essential testing ground for equipment and ideas before they were considered for squadron service. At this period AI Mk. II was under test, a number of sets also having been fitted to aircraft of Nos. 25 and 600 Squadrons. These early trials concluded:

> The effective field of AI vision was restricted to a fairly narrow cone directly in front of the aircraft and unless the fighter found himself between 6,000 and 1,000ft behind the enemy aircraft, and flying in the same direction at roughly the same height, he had very little chance of completing the interception.

They also highlighted the problems of providing adequate technical back-up at unit level, the early sets being very prone to failure. The training and provision of such groundcrew personnel was slow to develop, causing great frustration at the squadrons.

Improvements came with AI Mk. III by applying technology from the Air to Surface Vessel (ASV) Mk. I transmitter, thus increasing the maximum theoretical detection range to 17,000ft, but a host of other problems remained. In May the Mk. IIIA version was giving a maximum detection range of 9,000ft and a minimum of 900ft, but had a tendency to burn out when switched on. The Mk. IIIB, by using two aerials, achieved a minimum range of 600ft, which was much more in line with what was required; the minimum range was as important, if not more so, than the maximum range. Overall it was very frustrating, to say the least, and even on the occasions when an intercept did take place, the poor performance of the Blenheim usually led to the enemy aircraft escaping. Reluctantly, the decision was taken to slow down the employment of AI until some of the problems had been solved, rather than risk destroying all confidence in the equipment's potential. The only other technical improvement in May was the adoption of vertical rather than horizontal aerials, and this appeared to solve some of the ambiguity problems.

There was growing evidence that German aircraft were using some kind of radio beacon to aid navigation to their targets in Britain, and the NIC called for research into the nature of this equipment. The discovery of the 'Knickebein' beam, partly through excellent work on a number of Anson flights, was in many respects the true birth of the radio countermeasures (RCM) 'war' that was to play an ever-increasing role in the conduct of air warfare, and especially night warfare. July saw the formation of No. 80 Wing as a specialist, ground-based, RCM organisation to work in conjunction with the 'Y' Service (the Wireless Intelligence Service). Throughout the remainder of this account, various aspects of the RCM war will be examined as and when they affect the story.

On the night of 18 June the Luftwaffe launched its first night raid against England. It was to prove an enlightening experience for attacker and defender alike. The attack was to be made by Heinkel He 111s of KG4, based at Merville and Lille-Roubaix, and the primary targets were the airfields at Honington and Mildenhall, two of the bases from which Bomber Command was launching its nightly offensive against Germany. Chris Goss has researched this raid in British and German records, and the following account is a summary of his article, which appeared in *Victory Through Air Power*, a *FlyPast* special magazine published in April 1995.

> The lead pilot, Oberleutnant Ulrich Jordan, phoned his headquarters to ask about British night fighters and was told: 'Definitely not! We have no night fighters ourselves and I am sure that the British have none either. Don't

worry about meeting British fighters at night.' The first bombers were airborne about 2230 . . . meanwhile, among the British squadrons on standby that night were four in the southern part of the country – Nos. 19 Squadron (Duxford) and 74 Squadron (Rochford), both flying Spitfires, and Nos. 23 Squadron (Colleyweston) and 29 Squadron (Debden) with Blenheims.

As the German bombers approached the coast, the first fighters were already being vectored towards them. The Heinkel of Hauptmann Hermann Prochnow was the first to be lost, crashing into the sea near the Cork Light Vessel. However, the circumstances of this loss are not certain, although it is possible that this was the aircraft attacked by Flight Lieutenant 'Sailor' Malan of No. 74 Squadron, who reported a combat near Foulness Island. The next one was certainly down to him. Noting searchlight activity over Southend, he climbed to about 12,000ft and headed towards the area, sighting a Heinkel held in a searchlight beam. Manoeuvring for an attack: 'I gave it two five-second bursts and observed bullets entering all over the enemy aircraft with slight deflection as he was turning to port. Enemy aircraft emitted heavy smoke and I observed one parachute open very close . . . I followed him (the aircraft) until he crashed in flames near Chelmsford.'

The victim had been 5J+GA of the Staff flight, flown by Leutnant Erich Simon, on a diversionary raid against oil tanks at Southend. The next wave of bombers included that of Ulrich Jordan – and he was about to discover that, contrary to what he had been told, the RAF *did* have night fighters!

. . . a few minutes later, they were behind us – Blenheims! The combat lasted nearly twenty minutes, the fighters attacked continuously from behind, one after another, causing lots of damage to the fuselage, engines and tail of my plane, the seats being protected by armour plates saved my life. I ordered the bombs to be dropped and decided to try and make any airfield in German occupied territory. However, the fight continued until the other engine stopped, which happened just after the last fighter disappeared.

The two Blenheims of No. 23 Squadron, YP-S and YP-L, had both been hit by return fire from the bomber. The former aircraft (L1458) caught fire and Sergeant Close and his gunner, LAC Karasek, had to bale out; the aircraft crashed at Terrington St Clement, near King's Lynn. Meanwhile, Jordan's Heinkel had also caught fire and the pilot had to ditch near Blakeney Creek. The crew, including the badly wounded flight engineer, were taken prisoner by a Home Guard unit. Even as this Heinkel was going down, another was under attack by No. 23 Squadron. Squadron Leader O'Brien saw an aircraft, which turned out to be 5J+AM, held by a searchlight. As he moved into the attack, his Blenheim was spotted by the German crew – but neither aircraft had spotted the stalking Spitfire of Flying Officer 'Johnnie' Petre from No. 19 Squadron. The Spitfire and Blenheim opened fire at almost the same instant, but Petre then had to take avoiding action to miss the Blenheim. Although the Heinkel had received numerous hits, its gunners continued to fire at their attackers, shooting down the Spitfire when it broke away and was illuminated by a searchlight. Although Petre baled out he had been badly burned. Moments later the other two aircraft had also run into trouble – Willi Maier, unable to control the Heinkel, ordered the crew to bale out; however, when O'Brien manoeuvred to avoid the out-of-control bomber, his aircraft went into a spin, so he too ordered an abandon aircraft – from which only the pilot survived.

The events of the night were not yet over. North of Southend another pilot of No. 19 squadron, Flight Lieutenant Wilfred Clouston, saw an aircraft held by searchlights and went to investigate. It took him some five minutes to overhaul his prey – now identified as a Heinkel (5J+FP of 6/KG4). His two bursts hit both engines and damaged the electrical and hydraulic systems. The stricken bomber ditched near Cliftonville in the Thames Estuary. One crewman was killed but the other three, having taken to their dinghy, were made prisoner by the Home Guard. The final combat was probably between Blenheim L6636 of No. 29 Squadron and the aircraft of Leutnant Backhaus. The only certain fact is that both the Heinkel and the Blenheim crashed – the bomber on a beach near Calais and the Blenheim in the Thames Estuary. The balance sheet revealed a German loss of six aircraft (with six crewmen killed, three missing and eleven prisoners) and an RAF loss of three Blenheims and one Spitfire, with four aircrew killed.

In comparison to the scale and casualties of the daytime battle these were light, but the significance of the combats was far greater. The RAF's single-engined fighters had proved themselves capable, in co-operation with the searchlights, of engaging and destroying the night raiders. The Luftwaffe was no longer in any doubt that the RAF could, and would, contest the night skies over England.

June also saw the Luftwaffe begin night raids over other parts of the British Isles; small-scale bombing or reconnaissance missions to coastal areas and, on occasion, as far as Bristol and Mersyside. The searchlight organisation now comprised some 4,000 searchlights of various types, but there were still gaps in the planned zones. Nevertheless, they were at this stage the critical night asset, operating with fighters, as demonstrated by the June action detailed above, and guns. However, by the end of June it was concluded that:

> The success enjoyed by the searchlights had been short-lived. By flying at greater heights and by employing tactics of evasion, enemy aircraft had managed to avoid illumination. The fitting of AI to the Blenheims, however, held promise that an alternative system of operating against unilluminated targets could be developed – but the conditions to be satisfied before such interceptions could take place were formidable. The fighter had to be placed within several hundred yards of the enemy and on the same course.

By July 1940 the British radar chain comprised 30 CH and 30 CHL stations spread around the coast from Scapa Flow in the Orkneys to

Strumble Head in Wales. The associated Sector Operations Centres (SOCs), filter rooms, etc, were also in place – the system was, in theory, equally suitable for day or night defensive operations. Success came on the night of 22/23 July, when Flying Officer Ashfield, Pilot Officer Morris and Sergeant Leyland in an FIU Blenheim shot down a Dornier Do 17Z after performing a successful intercept on the bomber. The Dornier, from 2/KG3, crashed into the sea, and Leutnant Kahlfuss and his three crew were rescued.

Just before this, the FIU staff realised their limitations in the production of AI equipment, and it was agreed that a development contract for the next set, AI Mk. IV, be awarded to EMI. The company was to base its work upon new principles, especially the suppression of close-range clutter, in an effort to achieve a minimum range of the order of 400ft. Progress was rapid, and the FIU had the first sets back for trials in late June. They were greeted with great enthusiasm as, on first impressions, they appeared to be a huge improvement on what had gone before, and they did not interfere with the R/T.

In September it was decided to concentrate on production of Mk. IV sets, and to fit these to Bristol Beaufighters as a matter of priority. However, the slow build-up of the Beaufighter force led to the employment of six single-engined day fighter squadrons in the nightfighter role, three Defiant units (Nos. 73, 85 and 151 Squadrons) and three Hurricane units, this latter move not taking effect until October. Dowding was not impressed with having to adopt such makeshift measures: 'Our task will not be finished until we can locate, pursue, and shoot down the enemy in cloud by day and night, and AI must become like a gun sight,' he said.

As had been demonstrated by the success of the Spitfire squadrons in June, standard day fighters were able, given the right conditions, to adapt to the night scenario. They continued to be so employed over the next few months, but decreasingly so as the day battle grew more intense and they could not be spared.

While the value and potential of the AI-equipped twin-engined fighter had been realised, it was still early days in the development of the new technology, and great reliance had still to be placed upon the more traditional methods of airborne night defence and the employment of aircraft for radar-assisted visual acquisition. There were two basic techniques; the use of direct control via a D/F fixing system and the tried (First World War), tested and wasteful technique of standing patrol lines. In the latter method a section of aircraft (from one to three) would patrol a line delineated at each end by ground flares or searchlight marker beacons. Information on enemy aircraft positions was relayed by the sector controller, although this was only very general and almost total reliance was placed upon visual acquisition.

The Operational Record Book (ORB) of No. 29 Squadron for August 1940 reflects the nature of operations during this period:

> 8/8/40 One aircraft left Ternhill to patrol the Mersey Blue Line. There was considerable enemy activity but no bandits were illuminated. At 2340 an aircraft piloted by F/Sgt Munn left Ternhill on patrol, and at 2354 hours the Special Equipment Operator reported an aircraft at 5,000ft, about 4,000 yards to the left. F/Sgt Munn followed this aircraft for about 90 minutes until it went below 3,000ft in hilly country.
>
> 9/8/40 Several patrols went up during the night. The weather was generally unfavourable. Thick cloud hampered the searchlights, and although enemy aircraft were in the vicinity of our patrols, the special equipment failed to pick anything up.

Frustration mounted, both with the lack of performance of the aircraft and the disappointing results of the new equipment.

Thus, even among the AI-equipped units, the majority of action was still due to visual acquisition, often aided by searchlights or the gun barrage. On the night of 28/29 August Blenheim 'D' of No. 600 Squadron was scrambled:

> I was told to get off as soon as possible and to patrol base at Angels 17. For the next hour we received a good number of vectors and investigated innumerable searchlight concentrations. We were sent to patrol a line across which a lot of enemy aircraft were making their way to the Midlands. It was not long before I saw the exhaust flames of an aircraft close in front and above, so turned and went flat out after him, to find that we could hardly climb any higher, and all controls were pretty sloppy. So having staggered into a line astern position at approx 400 yards, let go a good burst. The enemy aircraft turned and dived, proving too fast for us to catch. A bit later I saw exhaust flames below us to starboard, so dived after them and was getting in really close when we were illuminated from behind. One searchlight coming in from the front, flicked over the aircraft in front before fastening on to us. It showed sufficient to show that we were very close, so opened fire before the searchlight blinded us entirely. Exhausted the remainder of my ammunition and again saw enemy aircraft dive away too fast for me to catch.

The Blenheim landed at Hornchurch at 2320. The only German casualty recorded for the early part of 29 August was Do 17Z-3 U5+PF from II/KG2, which crashed at Selis-Persan after receiving damage from an RAF nightfighter. However, there is no proof that this was a result of the above attack.

The Chief of Air Staff (CAS) called for a review of the night defensive system. This highlighted the shortage of twin-engined nightfighters and the need, therefore, to employ single-engined fighters in order to try and achieve a greater measure of success. Meanwhile, continued reliance had to be placed on the gun/searchlight combination for point defence of major cities and areas of industrial importance. In July these elements comprised 1,749 guns and 3,922 searchlights, manned by 157,319 personnel. The German night tactics were summarised thus:

> Night bombardment had changed its character towards the end of August when a greater attempt was made to concentrate attacks and damage definite areas, especially urban areas in which industry was centred. The Luftwaffe began to undertake a greater night assault.

Up to late August no urban centre night raids had involved more than 20 aircraft. From that time onwards the raids comprised up to 100 or more bombers. In August, Fighter Command operated defensive missions on 26 nights, flying 828 sorties but making only three claims.

In a speech delivered on 9 September 1940, Adolf Hitler decreed: 'We are giving our reply night after night. If the British declare that they will attack our cities heavily then we will wipe out their cities.' Even allowing for the heavy political rhetoric, the message was clear. The Luftwaffe would be ordered to intensify its night offensive.

Just a few nights earlier, on the 4th, the growing effectiveness of the RAF's night defences was ably demonstrated when Pilot Officer Herrick and Sergeant Pugh, in a Blenheim, shot down two enemy bombers, an He 111 and a Do 17: 'The pilot stated that the searchlights were most effective, and, of course, entirely responsible for enabling him to sight and fire at the enemy'. Merrick was awarded a DFC for this action; nine days later he claimed another He 111.

The growing night threat meant that an increase in fighter strength, and improved facilities and techniques, were essential before the winter operation period. There was no shortage of ideas. Among the night defence experiments under trial at this period were ALBINO and the Long Aerial Mines (LAM). The former was a version of the Parachute-and-Cable (PAC) system used to defend airfields from low-level attack, although in this instance it involved charges being attached to the cables of balloons. The latter entailed a bomber being targeted against one attacking aircraft track to lay a string of LAMs in front of the raider, the

LAMs consisting of an explosive charge on a cable, supported by a parachute. The idea was very simple. The enemy aircraft would have to fly through the line of LAMs and, hopefully, would hit one of the cables. This particular experiment was not abandoned until 1941, following the proven success of the third system, GCI. A Fighter Command memo of August 1940 highlighted the problems of using existing CH and CHL radar and control systems for night interception and then tasked the Telecommunications Research Establishment (TRE) and the Royal Aircraft Establishment (RAE) with developing new equipment and procedures.

The first experimental GCI station had been at Worth Matravers, under the TRE, and comprised a modified CHL radar for experiments with AI Mk. III-equipped Blenheims. The major improvements were to provide reasonably accurate height-finding and the use of a single Plan Position Indicator (PPI) to display both fighter and target, thus removing many of the previous problems of intercept geometry, the basic concept now being that of directing the fighter blip into a position behind the target blip using a curve-of-pursuit technique.

The first true GCI station was operational in October at Durrington, near Worthington, and for its first three months it remained under the control of Bomber Command's Operational Research Section (ORS) for trials. It soon became apparent that the main problem was that of capacity, with a maximum rate of three or four engagements an hour for each scope/controller. An interim solution was to use two controllers, each with one target/fighter combination, with a third controller plus assistants to carry out allocation of priorities, liaison with Sector Control and so on. Although it is essential to this story of nightfighting to cover the basic developments and techniques of GCI, it is not possible to mention all the modifications to equipment and changes of technique. At this critical stage in 1940 (and into 1941) the role of GCI was of particular importance, as the AI sets themselves still had severe acquisition limitations. As the war progressed, however, airborne radar became much more efficient and adaptable.

While searchlights and visual acquisition remained the primary means of detection, both sides gave careful consideration to the use of aircraft camouflage and other ways of making aircraft harder to spot. Black was still the favoured colour, for obvious reasons, but various shades and finishes were tried. The RAF's nightfighter scheme for 1940 was specified as RDM2, a velvety non-reflective surface that would, it was thought, absorb any light and help to make the aircraft invisible. Various reports were written over the years, either supporting or opposing this contention; it seems likely to be one of those areas of history in which absolute agreement will prove impossible. By October 1942 the RAF had virtually abandoned the idea of special camouflage in favour of the standard dark green/medium sea grey upper surfaces and sea grey undersurfaces. As with all aspects of camouflage, however, there were wide variations.

During the latter part of August 1940 the crews of a number of Defiant squadrons converted to the nightfighter role. The turret-equipped fighter had proved far too vulnerable in daylight, and the need to increase the night defences of England gave it a new lease of life. It was to be a 'cat's-eye' operation relying on visual acquisition of the target, and thus suffered from all the inherent problems mentioned above. It was not until 1941 that a number of Defiants acquired AI Mk. IV. The operational concept was for the Defiant, having acquired its target, to fly alongside and slightly below the enemy aircraft, allowing the pilot to keep station by silhouetting it against the night sky while the gunner put his four machine guns to work. Defiant N1621 of No. 264 Squadron, crewed by Pilot Officer Desmond Hughes and Sergeant Fred Gash, made such an intercept at 0155 on Wednesday 16 October 1940. The crew recorded the incident:

> It was a bright moonlight night. Suddenly, out of the corner of my eye I saw something move across the stars out to my left. If you are scanning the night sky it is normally completely still, so anything that moves attracts the eye. This just had to be another aircraft. I got Fred to swing his turret around and we both caught sight of a row of exhausts. It was a twin-engined aircraft. I slid alongside, below and to the right of him, and slowly edged in 'under his armpit' while Fred kept his guns trained on the aircraft. Then we saw the distinctive wing and tail shape of a Heinkel – there was no mistaking it. I moved into a firing position, within about 50 yards of his wing tip and slightly below, so that Fred could align his guns for an upward shot at about 20 degrees; obviously the German crew had not seen us, they continued straight ahead.
>
> Fred fired straight into the starboard engine. One round in six was a tracer, but what told us that we were hitting the Heinkel was the glitter of de Wilde rounds as they ignited on impact. Fred fired, realigned, fired again. He got off two or three bursts. There was no return fire from the bomber – indeed, I doubt if any guns could be brought to bear on our position on its beam. The engine burst into flames, then the Heinkel rolled on its back, went down steeply and crashed into a field near Brentwood.

The aircraft was He 111H-4 1T+BB (or 1T+LK) of 2/KGr.126, and it crashed at Creasey's Farm at 0200 with the loss of two crewmembers (Unteroffizier Glaser and Gefreiter Tordik), the other two (Leutnant

Right, upper: Savy airfield, France, in October 1917. An R.E.8 of No. 3 Squadron, Australian Flying Corps, Starts up for a night bombing sortie. In the absence of an effective nightfighter force, the Germans relied on AA guns to counter night raids.
Right, lower: Sopwith Pup A6231 of No. 73 Squadron at Lilbourne, 1917. The Pup, like the Camel, was a fighter type that the authorities thought unsuited to night operations. However, some were operated successfully by RNAS units.
Below: A Zeppelin hangar at Epich. Sheds such as this became targets for night intruder operations as part of the overall campaign to defeat the bombers they housed.

Right, upper: Lieutenant Fred Sowrey of No. 39 Squadron in B.E.2c 4112, the aircraft in which he destroyed L32 on the night of 23/24 September 1916.
Right, lower: The S.E.5a did not see a great deal of service in the nightfighter role, although No. 61 Squadron did operate the type out of Rochford, Kent, in early 1918.
Below: A superb illustration produced by No. 50 Squadron, summarising that unit's part in the night defence of Britain. All of the elements of the night war are here; a flaming Zeppelin, a Gotha caught by one of the many searchlights, a fighter crashed on the airfield, and another taking off in the beam of a searchlight.

Right, top: The 'Zeppelin Killers' of No. 39 Squadron, Sutton's Farm. The middle three in the back row are Lieutenants W. Tempest, W. Leefe Robinson and Lieutenant F. Sowrey.
Right, centre: Bekesbourne, home of No. 50 Squadron and its B.E.12 nightfighters.
Right, bottom: The skeletal remains of L33. The Zeppelin fell on the night of 23/24 September 1916 to damage caused by AA guns and Alfred de B. Brandon of No. 39 Squadron.

Above: L32 is dismantled at Billericay, Essex, after being brought down by Lieutenant Frederick Sowrey.

Left, upper: In October 1917 No. 65 Squadron moved to France to fly defensive patrols and bomber escort. Like many fighter squadrons it was called on from time to time to operate at night.

Left, lower: Sopwith Camel B2402 of No. 44 Squadron at Hainault Farm in Essex. Under the dynamic leadership of Captain Gilbert Murlis Green, this unit pioneered the use of this day fighter in the nightfighter role, the first sorties being flown on 3/4 September 1917.

Above: The B.E.12, along with the B.E.2c, was the standard RFC nightfighter for much of the First World War.
Right, upper: A graphic demonstration of the use of underwing flares being used to assist pilots in navigation and landing.
Right, lower: RFC personnel inspected much German Air Force equipment at the end of the war. Among the fighter types was the Fokker D.VII.

Left: William Leefe Robinson, No. 39 Squadron's first airship destroyer.
Below: Bristol Fighters were among the types eventually employed to counter the attacks on London by advanced Zeppelins and large aeroplanes.

Right: The fate of many nightfighters on return to base. This B.E.2c is badly damaged after a landing accident. The majority of casualties among the nightfighter forces of both sides were the result of such accidents, which often had fatal consequences.

Above: Two of the German giants; Staaken R.VIs 36/16 and 33/16, both of Rfa 500.
Left: One of No. 39 Squadron's Bristol Fighters, with added firepower in the form of twin guns for the pilot and observer.

Right: The F.E.2b saw extensive employment on Home Defence duties.

Left: 2nd Lieutenant Cadzow, Home Defence pilot, in 1917.
Below: A B.E.2c in a night scheme. When used for reconnaissance on the Western Front the aircraft was a disaster, but its inherent stability proved advantageous when it was used as a nightfighter.
Right, top: Air doctrine in the 1920s favoured the bomber as the primary weapon and suggested that bombers would always be able to find and destroy their targets. Little consideration was given to night offence or defence. This Vickers Vimy is typical of the period.

Above: A Fairey Hendon night bomber of No. 38 Squadron, the RAF's most up-to-date bomber in the mid-1930s.
Left: A graphic view of a mock night attack on London in July 1927. A bomber of No. 99 Squadron braves the searchlights and fighters above a target that is fully lit by a myriad street and house lights.

Right, upper: It was essential to stop German bombers reaching their targets; the transfer of bomber squadrons from day to night operations in late September 1940 meant that the nightfighter force had to effect rapid improvements.
Right, lower: The Blenheim I was equipped with AI, and rapidly proved that the principles of the equipment were sound.
Below: A Blenheim IV of No. 143 Squadron.

Left: At Catterick in 1940, 'A' Flight of No. 219 Squadron pose in front of one of thier Blenheims.

Right, top: One of the first AI-equipped units, No. 25 Squadron undertook much pioneering work. This Blenheim is seen at Debden in late 1940.
Right, centre: A Blenheim IF of No. 25 Squadron at Debden in 1940. (Paddy Porter)
Right, bottom: Beaufighter VI KW103 in No. 406 Squadron markings. This was the RCAF's first nightfighter unit, having formed at Acklington in May 1941.

Right: 'A' Flight of No. 219 Squadron at Tangmere in 1941.
Below: Of the single-seat nightfighters, the Hurricane proved the most adept at the role. The aircraft were used extensively in the 'cats-eye' technique in 1941, No. 87 Squadron, as seen here, being particularly successful.
Bottom: An essential element of both day and night defences, the Sector Operations Centres were the heart of Fighter Command's system.

Right, top: In an endeavour to increase the interception rate, the 'hunter-killer' system of Turbinlite Havoc/Boston with a satellite Hurricane fighter was introduced.
Right, centre: The AI-equipped Turbinlite aircraft (AH470 is shown here) controlled the nightfighter formation under instructions from GCI.
Right, bottom: The TRE at Defford became a centre of AI radar development. This 'thimble-nosed' Beaufighter is being used for trials.

Left: A No. 140 Squadron Beaufighter with the early 'bow-and-arrow' AI aerials.
Below: Beaufighter IF mod, X7578.
Bottom: Defiant II AA370 with AI aerials mounted on its starboard wing, August 1941. A number of Defiants were given AI as and when suitable sets became available.

Above: The Beaufighter added a new dimension to the nightfighter capability of the RAF, both in performance and firepower.
Left, top: By 1941 most nightfighter units had re-equipped with the Beaufighter, and it was this type that began offensive night operations using the 'Serrate' equipment.
Left, centre: A Ju 88C, probably of I/NJG2. The type proved to be a remarkably efficient nightfighter, and also operated in the night intruder role, a mission instituted by NJG2.
Left, bottom: A trio of Bf 110 nightfighters. The prominent nose aerials adversely affected on the aircraft's performance, but in the early years, when there was no airborne fighter opposition, this mattered little.

Above left: A Messerschmitt Bf 110 of II/NJG4 with SN2 at St Dizier in 1943. (via Chris Goss)

Above right: Ju 88 variants became powerful and effective nightfighters.

Below: His Majesty King George VI meets 'A' Flight of No. 257 Squadron at Honiley. The Squadron's Hurricanes were acting as satellite fighters for the Turbinlite aircraft of No. 1456 Flight.

Newald and Gerfreiter Granetz) having baled out to become POWs. The wreck of the Heinkel was excavated in 1978.

Landing from the night's operation, the Defiant crew overshot the runway in poor weather conditions, although without injury and with only minor, repairable damage to N1621.

The Luftwaffe night effort on the 15/16th had been heavy, with approximately 400 sorties flown. Some crews flew two missions that night, and the standing order was for aircraft to stay over their targets for as long as possible and 'nuisance bomb' every few minutes. The RAF responded with 41 sorties, but only two resulted in intercepts, one by an AI-equipped Blenheim of No. 23 Squadron which was unable to achieve a firing position, and the other the successful engagement described above. The raid caused heavy damage to London, some 900 fires, many of them classed as major, being reported. The poor showing of the defences caused serious questions to be asked within Fighter Command; there were still those who doubted the use of AI and considered that resources should be used elsewhere.

John Cunningham was destined to become one of the great nightfighter aces, but his early experiences with the AI aircraft caused serious doubts:

> We had vectors from Sector Control and were then instructed to 'flash your weapon'. The AI op was unable to pick up the target, then I saw the bombs go down. The magician was still kneeling on his prayer mat of blankets muttering to himself, the green glow from the CRT flickered on his face. A witch doctor, I thought, a witch doctor and black magic – and just about as useful.

These were, indeed, difficult days.

During September, Sir John Salmond headed a committee examining all aspects of the night defences, and among a long list of recommendations and proposals suggested that:

1. Production of Mk. IV AI should be accelerated along with its fitting to Beaufighter aircraft of Fighter Command.
2. GCI should be provided from coastal radar stations.
3. Additional radio aids should be provided.
4. Specialised nightfighter training should be provided for crews, probably at a nightfighter OTU.

On the 21st of the month Dowding circulated a memorandum entitled 'Night Interception', which stated:

> The basic principle is that one searchlight per section (approximately one in six) shall be equipped with the GL [gun laying] set or some other radio

equipment which will enable the position of an aeroplane (including its height) to be determined.

The present difficulties in the way of night interception, even with AI, are that the tracks resulting from Observer Corps plots are extremely spasmodic and inaccurate. This is quite understandable, because enemy aircraft now generally fly at great altitudes and almost never within sight of the ground. Most encouraging results have been obtained from the GL sets already installed in the Kenley Sector; they have proved more capable of maintaining continuous tracks than have the RDF stations on the South Coast, which are still suffering from the after-effects of bomb attacks.

The technique would be that the Sector Controller would make up his mind that such-and-such a raid offered a good opportunity for interception. He would give the order to concentrate on that raid and, thereafter, no searchlight would be permitted to expose any other echo. Even on the selected raid no searchlight other than the 'master' searchlights would be permitted to expose.

The first master searchlight to pick up the echo would determine the height of the aircraft and at once pass it by direct land-line to Sector HQ. The Sector Controller would thereupon pass the enemy's height to the fighter on patrol. The Sector Controller would then give vectors to his fighter to bring him into contact with the head of the enemy's track. On a clear night, therefore, the Controller's task is simply to direct his patrolling fighter so that it can see a single searchlight beam. If the fighter pilot sees a single beam he will know that it is pointing at the target and, having been given the height, he knows exactly where to look.

If he sees two beams converging then he will know that two master beams have opened up and that the enemy is to be found at the point of their convergence. The point which we must discover by experiment is how accurately the searchlight radio sets can determine the height of the enemy. If this can be done with approximate exactitude the method, on a clear night, should enable any fighter to make an interception, whether fitted with AI or not. When the fighter patrol is working above a continuous cloud floor, the Controller's vectors, supported by the pool of light on the upper surface of the cloud, should enable a fighter fitted with AI to make interceptions on a good percentage of occasions.

There are indications that even the superior performance of AI Mk. IV set will not result in a high proportion of interceptions by methods hitherto employed. The special features which give promise that the above-described procedure will give better results than any system hitherto tried are:

1. The height will be approximately known.

2. The fighter pilot will know that whenever he sees a searchlight beam it is pointing directly or almost directly at the target.

3. The tracking of the bomber's course on the Sector Operations Table will be very much improved.

> Other refinements, such as the fitting of a Lorenz set to take advantage of the enemy's beams, may be later introduced, but I wish to avoid complicating the initial trials by non-essential adjuncts.

September 1940 saw the Battle of Britain at its height and the RAF's day fighters' defeat of the Luftwaffe's attempts to subdue British defences through daytime bombing. London had suffered heavy damage during the latter stages of the battle and, as Dowding had realised months before, with the end of the daylight campaign the German bombers would resort to ever heavier night attacks. Great improvements had been made to the night defences of southern England, but would they stand up to the test?

The night of 7/8 September brought a major raid against London, some 40 aircraft passing in three waves over the city at about 2100. Bombs fell on Battersea, Paddington and Hammersmith, the main attacks being followed by single 'nuisance' raiders at regular intervals to add to the disruption and prevent firefighting. Among the defenders were three Blenheims of No. 25 Squadron and two from the FIU, the latter reporting: 'Numerous AI contacts were obtained but constant interference from undirected AA fire and searchlights prevented success.' During this phase, Dowding commented: 'The most depressing factor which has emerged from the past weeks is that the Germans can fly and bomb with considerable accuracy in weather in which our fighters cannot leave the ground'.

The role played by the anti-aircraft guns remained of critical importance, not only for the destruction of enemy bombers but also because of the effect of the barrage on morale. Nevertheless, it also interfered with nightfighter operations despite attempts at imposing 'no fire' zones and height bands.

As part of the expansion of the RAF's nightfighter force, Flight Lieutenant J. Sanders, ex-FIU, was given command of the newly-formed 422 Flight on 14 October, the unit being equipped with Hurricanes for night defence. A few days later, on 25 October, the Italian bomber force made its first night attack on the British mainland when sixteen Fiat BR.20s attacked Harwich. One crashed on take-off and the others, having caused almost no damage, had problems finding their airfield. Two ran out of fuel and crashed, the crews having baled out. It was not an auspicious start. That same night saw one of the few AI successes when a Beaufighter of No. 219 Squadron claimed a Do 17. The squadron had flown 22 patrols during the last two weeks of October, during which they recorded 29 AI contacts, leading to 10 visual pick-ups and four engagements. The statistics were improving, but even though the number of AI contacts had increased, all too few resulted in combats, the problem being, in part, the limited performance of the Blenheim.

In October the decision was taken to employ three squadrons of single-engined aircraft in the nightfighter role. Numbers 73, 85 and 151 were chosen, although No. 73 was then replaced by No. 87 Squadron, but this employment was not intended to be restrictive:

> To relegate any single seat aircraft wholly to night duty was dangerous and unsound . . . every fighter should be capable of leaving and returning to its aerodrome blind. Our task will not be finished until we can locate, pursue and shoot down the enemy in cloud by day and by night, and the AI must become the gun sight . . . nothing less will suffice for the defence of the country.

Six basic techniques were proposed:

1. Fixed patrol lines.
2. GCI overland control – using data from Observer Corps, not radar.
3. GCI seaward – using radar.
4. Predicted coast-in point defence.
5. AI aircraft with radar-directed searchlights.
6. Aircraft with Lorenz receivers to hunt along German navigation beams.

One of the primary sources for details of Fighter Command operations is the 'Form Y' summary. Typical of these documents are those for the nights of 13/14 and 14/15 November. Together they illustrate the position of the defences during the latter part of 1940:

Night 13/14 November.

Night fighter patrols.

9, 10, 11 Groups	Nil
12 Group	Wittering – Blenheim patrol Crowland/Sutton Bridge 8,000ft at 1855, investigate raid at 1909, intercept raid at 1941. Blenheim patrol Bircham Newton 2030, intercept raid 2040. Kirton – Blenheim patrol Kirton 7,000ft at 2037, patrol Scampton at 2050.
13 Group	Turnhouse – two Blenheims and one Hurricane. Usworth – 3 Spitfires. Catterick – 3 Blenheims.
14 Group	1 Blenheim.

There were no intercepts, although 70–80 enemy aircraft active.

Night 14/15 November.

Night fighter patrols.

9 Group	Nil
10 Group	34 patrols, including 5 Gladiator.
11 Group	49 patrols – 26 Defiant, 7 Hurricane, 6 Blenheim, 10 Beaufighter. 5 e/a seen but no intercepts.
12 Group	24 patrols. Blenheim from Digby claim one e/a dam aged near Swaffham.
13 Group	10 patrols.
14 Group	3 patrols by Hurricanes.

. . . enemy operations have been on a large scale and it is estimated that at least 350 e/a have operated, the majority of which flew at 12–15,000ft. Coventry was the main target and was heavily bombed from 1900 hours by approximately 280 aircraft.

November 19/20 saw the first Beaufighter kill, the destruction of a Ju 88 near Brize Norton, the aircraft falling to a No. 604 Squadron aircraft flown by John Cunningham and Sergeant J. Phillipson. The squadrons had only been given Beaufighters in October (Nos. 25 and 29 Squadrons being the first to acquire the type) as part of the planned replacement of the slower, less manoeuvrable Blenheims. The Bristol Beaufighter had its origins in 1938, being conceived as a fast, well-armed fighter. The obvious potential of the airframe, and the need for an improved nightfighter capable of being fitted with AI, led to its employment in the night role.

Taken overall, the statistics given above were not very reassuring for the defenders, and confirmed the earlier opinion that the effectiveness of the searchlight/day fighter combination was all but ended. More AI-equipped aircraft were needed, plus the trained crews to operate them.

Despite the 'priority' given to installing AI in Beaufighters, by November only 47 aircraft had been so equipped. Progress was slow but steady in the vital area of training, and December saw two important moves – the formation of No. 54 (Night) Operational Training Unit (OTU) at Church Fenton, and the decision at No. 3 Radio School, Prestwick, 'to provide operators with air training before they arrived on the squadrons'. This referred to the Radio Operator (Air) category. It was realised how vital they were to the success of the technique:

It is of the greatest importance that the AI operator should be intelligent, keen, and of a patient and painstaking disposition. He is, after all, the brains

> of the aircraft up to the moment when the pilot actually sees the silhouette of the enemy aeroplane and opens fire.

In mid-December No. 422 Flight had been expanded to become No. 96 Squadron, and on the night of the 22nd of that month Flying Officer P. Rabone was on patrol over the Formby area at 14,000ft in Hurricane V6887:

> . . . he sighted an unidentified enemy aircraft some 50 yards ahead, 25ft above and opened fire into the belly of the aircraft, which nosed down in front of him, and as the enemy aircraft passed he followed it in his sights and managed two more bursts. As he flew above the enemy aircraft through its track, he passed through a stream of oily smoke. The enemy aircraft was first sighted because of his exhausts.

The 'cat's-eye' pilots were still doing the most damage, despite their obvious limitations.

Since the raid on Coventry in mid-November, German bombers had concentrated on industry and ports, attacking Coventry, Liverpool, Southampton, Manchester and Sheffield, with London as the secondary objective. Using pathfinder techniques by KGr.100, the Luftwaffe were achieving good concentrations of effort and inflicting serious damage. The Coventry raid of 29/30 December was described thus: 'For the first time air power was massively applied against a city of small proportions with the objective of ensuring its obliteration.' The main raid covered the period 2020–0610, and some 437 aircraft took part. Although Fighter Command was expecting such an operation, and under Operation Cold Water had conceived a defensive counter-plan, the 119 patrols met with no success.

An option that had been under consideration for some time was that of taking the night war to the enemy. In December, No. 23 Squadron were ordered to remove the AI sets from eight aircraft (for security reasons, in view of the danger of losing an aircraft over enemy territory) and to have these standing by for a special mission over French territory.

The earliest such intruder patrols had been flown in August, using Blenheims provided by 2 Group (although No. 604 Squadron had operated over the Pas de Calais airfields in May and June). However, this had been shortlived, as Bomber Command's resources were limited and the primary aim of that organisation was to expand and improve its bomber force for the offensive against Germany; thus, Fighter Command took on these 'security patrols'. The initial plan of August 1940 called for one flight of No. 23 Squadron to specialise in the intruder role, 'to attack enemy bombers in the vicinity of their aerodromes, and to attack with machine gun fire aircraft and personnel on the ground'. It was the lack of Bomber Command involvement that led to the decision to employ the whole of No. 23 Squadron for the new role. Information from the Y Service was used to determine the most appropriate targets for attack on any given night, the operational area being divided into three zones:

Group 1 Lille-Nord, Lille-Vendeville, Vitry en Artois, Cambrai-Epigny, Cambrai-Niergnies.
Group 2 Amiens-Glisy, Poix, Beauvais-Tille, Montdidier, Rosierres-en-Santerre.
Group 3 Evreaux-St Martin, Evreux-le-Coudray, Dreux, St André-de-l'Eure, Caen-Carpiquet.

The operation was delayed because of poor weather over France, but at last, on the evening of 21 December, six aircraft took off to carry out offensive patrols over the German airfields in the Abbeville, Amiens and Poix sectors. This was also the night that the Luftwaffe mounted a major attack on Liverpool. The first enemy bombers were detected at 1700, and the first bombs fell in the target area just under two hours later. Warned to stand by for intruder operations, No. 23 Squadron sent three aircraft to attack selected airfields. Typical of the night's missions was that by Flying Officer Williams and crew, who departed Ford at 2020. The Blenheim went first to Poix, where it circled for 20min, dropping a number of 20lb bombs, at which point the airfield lights were doused. Williams then moved on to do the same at Montdidier and Abbeville. The RAF account of this first operation concluded:

> The presence of our aircraft could hardly fail to cause some disturbance to the enemy arrangements for homing and landing his bombers. This in turn was likely to raise the enemy accident rate and lower the subsequent serviceability of his bomber forces.

The intruder campaign that was to become a dominant feature of No. 23 Squadron's war was under way.

By the close of the year all the basic elements of the RAF's night defences were in place. In December 1940, Dowding, as AOC Fighter Command, summarised the problems: 'I am convinced that the main obstacle to night interception is the lack of accurate tracking inland from the coast, and most importantly of all, lack of accurate information with regard to the height of the enemy bomber.' In the same report he called for a specialist nightfighter force of twenty squadrons and an expansion of the other

TABLE 1. LUFTWAFFE NIGHT BOMBING CAMPAIGN OVER THE UNITED KINGDOM, JUNE–DECEMBER 1940

Month	Sorties	Claims		
		NF	AA	Other
June	1,367	16	–	–
July	1,527	3	4	2
August	4,050	3	16	4
September	6,950	4	31	3
October	6,520	3	21	7
November	6,025	2	20	5
December	3,040	4	10	–
Totals	**29,479**	**35**	**102**	**21**

elements of the night air defence network, including the provision of more radar-laid anti-aircraft guns. The following year was to see great progress in all these areas, although by then the threat to the British mainland was minimal. The same could not be said of the bomber offensive aimed at Germany. This had begun in 1940 and in response the Luftwaffe instituted a rapid expansion of its home defence forces.

For the period June to December 1940, RAF records show a total Luftwaffe night effort against the United Kingdom of 30,379 sorties. The defences claimed 137 aircraft, with a further 21 being lost to 'other causes'. The detailed breakdown of these statistics is shown in Table 1.

The year 1940 was also one of development for the Luftwaffe, as the weight of Bomber Command attacks shifted to the night hours and the Strategic Bombing Offensive truly began. During January 1940 Wellingtons and Hampdens had joined the Whitleys in the night task of leaflet-dropping over Germany, primarily to allow crews to gain experience of night operations rather than being due to any desire to increase the amount of paper being spread over Europe. The policy of dropping propaganda leaflets (codenamed 'nickelling') caused questions from many quarters as to its value. Certainly the leaflets themselves were most unlikely to have any effect on their recipients (other than to provide an excellent source of paper for certain uses). However, while bombing of the mainland remained prohibited there was little else that could be done to give crews this invaluable experience against the day when the missions would take on a more offensive nature.

On the night of 19 March 1940 a force of 30 Whitleys and 20 Hampdens attacked the German seaplane base at Hornum on the southern end of the island of Sylt. The significance of this attack lay in its being the first bombing mission against a land target in Germany (in reprisal for a German attack on Scapa Flow). Although the bombers claimed to have identified and attacked the targets, subsequent reconnaissance showed no damage.* There had been virtually no opposition to the bombers; certainly no fighter activity. During that same month Bomber Command's Operational Research Section issued a report on the loss rates for the period September 1939 to March 1940, showing an average of 13 per cent for daylight operations, compared with 2 per cent for night operations. Although Bomber Command never completely abandoned daytime operations, it is true that from spring 1940 onwards the emphasis moved towards night operations.

Events now moved rapidly to bring into play the factors that would dominate the air war in Europe for the remainder of the war. In early April the Germans invaded Denmark and Norway, eliciting a weak response from the British military. Then, on 10 May, came the long-awaited offensive in the West as German forces steamrollered into the Low Countries and France. As early as 13 April a new bombing directive had been issued to the RAF, stating that, in the event of general air action being called for, Bomber Command was to implement WA 8 (Western Air Plan No. 8), the night attack on Germany, the priorities being given as:

1. Oil installations.
2. Electricity plants, coking plants and gas ovens.
3. Self-illuminating objects vulnerable to air attack.
4. Main German ports in the Baltic.

A full-scale bombing offensive against German industry meant that the defences of the Reich would have to be modified to meet the threat. On 14 May the Luftwaffe launched a major air assault on the Dutch city of Rotterdam, the first overt bombing of a city by either side in the western theatre of operations. To many in the British War Cabinet this was looked upon as the final straw; the 'gloves were off' and the plans for an all-out bombing offensive could now be implemented. The following day, Churchill authorised the bombing of targets east of the Rhine. That same night, 15 May, 99 RAF bombers attacked oil and rail targets in the Ruhr and the Strategic Bombing Offensive (SBO) was under way. The Air Staff decided that the maximum impact could be derived from a concentration of effort against industrial targets in Germany, 'to cause the continuous interrup-

*The problems of target acquisition are beyond the scope of this book, but a good account of Bomber Command's war can be found in *The Six Year Offensive* by Ken Delve and Peter Jacobs, Arms and Armour Press, 1992.

tion and dislocation of industry, particularly where the German aircraft industry is concentrated'. As explained previously, it is not the intention here to follow all the nuances of the SBO, but is important to highlight strategic decisions and particular raids, as these had a direct impact on the development of the night defences.

German air strategy was based on offensive operations; how would it react to a concerted bombing campaign? At the outbreak of war the German night defences were virtually non-existent, comprising a limited number of searchlights and anti-aircraft guns (Flak – from Fliegerabwehrkanonen). The problem of locating and finding an enemy aircraft over Germany was no different from that faced by the British defenders, although the latter were somewhat more advanced with the development of ground-based radar and the integrated system of filtering and control. In February 1939 Luftwaffe orders of battle record seven nightfighter units (all with single-engined aircraft), but in view of the requirements of the air campaign of 1939 they were, by the early summer, absorbed into the day fighter units. Part of the rationale was that many still considered that day fighters should be able to operate satisfactorily at night, using a combination of searchlight techniques and whatever moonlight was available.

According to most records, the original nightfighter unit had been the Bf 109-equipped 'Lehrgeschwader' (LG2) at Greifswald, which was trained in searchlight co-operation for night air defence. After these early trials the unit was used as the basis of 10/JG26 under Johannes Steinhoff.

In September 1939 the fighter force comprised five Staffeln that were designated day/night: 11(N)/LG2, 10(N)/JG72, 11(N)/JG72, 10(N)/JG2 and 10(N)/JG26, all of which were equipped with Bf 109 variants; by December the organisation had been rationalised, the other units being merged into IV(N)/JG2 under Hauptmann Blumensaat. However, the Bf 109 was not suited to night operations, and the chances of the pilots achieving any worthwhile results were all but eliminated when they were ordered to operate only on cloudy nights, the clear, moonlit nights being given over to the searchlight/gun combination. Later in the war, when Bf 109 and Fw 190 variants were being used as 'cat's-eye' nightfighters, many pilots favoured the Bf 109 for the simple reason that its undercarriage would break cleanly in the event of a bad landing, leaving the aircraft on its belly and thus the right way up. The Fw 190, on the other hand, with its sturdier undercarriage, was likely to flip on to its back, trapping the pilot.

A posting to a nightfighter unit was invariably looked upon by the aircrew as a punishment post, and most day fighter commanders were certainly not keen to release their best men. For those on the nightfighter units, this early period was seen as sheer frustration: 'It was the same old story which had been going on for weeks; the alert would be sounded, the fighters would take off towards the threatened area, then invariably fail to make contact with the enemy.' The situation was not helped when the first victim of a Bf 110 nightfighter turned out to be another Bf 110.

The small scale of Bomber Command 'attacks' (with leaflets) during the period up to spring 1940 did little to encourage the creation of an effective night defence system. Nonetheless, it highlighted the difficulty of shooting down the enemy aircraft. As we have seen, Bomber Command losses were insignificant during this period, and it is likely that most were due to other causes (bad weather, especially icing, running out of fuel, and so on), rather than good shooting by the German gunners. Nightfighter successes were few and far between – some sources record the only confirmed kill as being that by Major Paul Förster of IV/JG2 in a Bf 109. Other sources list the first victory as that by Oberfeldwebel Willi Schmale against a No. 218 Squadron Fairey Battle on 20/21 April. Battle P2201 was on a reconnaissance and leaflet-dropping mission to Darmstadt and Mainz, and only the pilot, Pilot Officer H. Wardle, survived to be taken prisoner.

However, this attitude to the threat was to change with the bombing of land targets in the Reich. The initial impetus was not so much a fear of the material damage, but rather the loss of face to the Nazi regime, especially with the likes of the Luftwaffe commander, Reichsmarschall Hermann Goering making such bold statements as 'No enemy aircraft will fly over the Reich territory'.

Spring 1940 had seen the introduction of the Bf 110 to the nightfighter role following trials by I/ZG1. An aircraft that had proved a failure in its much-vaunted daylight 'Zerstörer' (destroyer) role was about to become a significant element in the night defence of the Reich. The trials were based at Düsseldorf, and the unit acquired the provisional title of 'Nacht und Versuchs Staffel' (Night and Experimental Squadron). The initial trials in the early part of the year had proceeded under the direction of Wolfgang Falck, and had involved the fighter orbiting a predetermined position until observing a target held by searchlights; overall co-ordination was by a ground controller. Although there had been no success, Falck and many others were convinced that this was the way forward. In common with the British experience it was realised that the twin-engined, two/

three-crewed fighter could provide the most effective solution, especially with the promised advent of airborne detection aids. As one would imagine, the experiences of the two opposing air forces and the impetus for development was very similar, the pace of change and the tactical approach adopted depending upon available technology, Command support and the exact nature of the threat.

By June the Luftwaffe had decided upon the Bf 110 as the backbone of its nightfighter force. Although many favoured the adoption of the Ju 88 with its improved performance, this aircraft was not available, whereas the Bf 110 was in need of a new role. Elements of I/ZG1 and IV/JG2 undertook night flying training in preparation for their new task. On 20 July 1940 ZG1 was redesignated NJG1, with its HQ at Ziese and operating from bases in Holland as part of Luftflotte 2. This was a result of a directive from Goering on 26 June, ordering Wolfgang Falck to form NJG1 as the first true nightfighter unit under the command of Luftflotte 2. By July, Falck's new unit had absorbed 2/ZG76, 3/ZG76 and part of KG30, the latter unit having the Ju 88 rather than the Bf 110. However, the KG30 element soon moved from Gütersloh to Düsseldorf as 4/NJG1 for intensive training, at the same time acquiring an additional Staffel equipped with special Dornier Do 17Z-10 'Screech Owl' aircraft. These had an infra-red (IR) detector capable of picking up engine exhaust emissions and displaying information on a 'Q tube' in the cockpit. These 'Spanner Anlage' systems

MESSERSCHMITT Bf110

Considered a failure in its much-vaunted role as a day bomber destroyer, the Bf 110 went on to have a fine career as a nightfighter, becoming the mainstay of the Luftwaffe's nightfighter units for much of the war. The prototype day fighter first flew in May 1936, and seemed to have great promise. However, in the 1940 combats with RAF aircraft the Bf 110 proved a dismal failure. In July that same year the Luftwaffe realised that it needed a specialised nightfighter, and the Bf 110 seemed able to fit the requirement. It entered this role with NJG1, although it was to be a year before the first AI sets were available. From that point the Bf 110 proved invaluable as a nightfighter, and over the next few years it appeared in a number of variants with changes of radar or weapon system. The type was still in service at the end of the war, although by that time the Ju 88 and He 219 were considered to be superior in the nightfighter role. The Bf 110G-4 was the variant that, during 1943 and 1944, bore the brunt of the night war, achieving marked success against the compact waves of RAF bombers.

Bf110G data
Powerplant: Two Daimler-Benz DB 605B-1. Crew: Three. Length: 41ft 6in. Wingspan: 53ft 4in. Maximum speed: 350mph. Ceiling: 33,000ft. Armament: A wide variety of cannon and machine-guns.

were one attempt at trying to turn night into day and, at shorter ranges, did prove of some use. Once the unit had completed its work-up it moved to Schiphol airfield near Amsterdam.

Progress was now quite rapid, and on 17 July Oberst Josef Kammhuber was given command of the newly created 1st Night Fighter Division, part of Luftflotte 2 under General Kesselring. By October Kammhuber was a Major-General as 'General of Night Fighters', responsible for defensive and intruder operations.

Oberleutnant Werner Streib scored the first success under the new organisation with the shooting down of a Whitley on 20/21 July. This night saw Bomber Command's biggest offensive effort since the fall of France, 81 bombers attacking targets in Germany. At 0200 Streib and Corporal Lingen were patrolling in their nightfighter when they spotted an aircraft some 200 yards to one side. At first they took it for an aircraft similar to their own, but moved closer to be sure. First they noticed that it had a turret, and then they saw the RAF roundel – a Whitley. The bomber's rear gunner fired a burst at the nightfighter, but with no effect. Streib replied with two bursts of fire:

> His starboard engine was burning mildly. Two dots detached themselves, two parachutes opened and disappeared into the night. The bomber turned on to a reciprocal course and tried to get away, but the plume of smoke from his engine was still clearly visible. I attacked again, aiming at the port engine and wing, without this time meeting any counter fire. Two more bursts and engine and wing immediately blazed up. The aircraft turned over and crashed into the ground.

This victory was achieved flying the one and only trials Dornier Do 17Z-7, a type that was being promoted as a potential nightfighter to overcome some of the drawbacks of the Bf 110. A small batch of the Z-10 variant entered service with I/NJG2, but the type never became a major element of the nightfighter force. The total of all variants built was approximately 340 aircraft. Two days later Streib claimed another Whitley, and added two Wellingtons to his score in August.

The basic technique was still that of the 'cat's eye', the so-called 'Helle Nachtjagd' (illuminated night fighting), whereby the roving fighter pilot relied on the searchlights to reveal his prey and then on the eyes of his crew to achieve the kill. The system initially had three zones in the neighbourhood of the Zuider Zee and Rhine estuary, thus covering the direct route to and from the Ruhr. The area covered was 90km long by 20km deep. The roving nightfighters were not permitted to stray over flak zones, as it was still thought that flak had the best chance of bringing down bombers over the targets. The nightfighter squadrons were, at first, moved from base to base to cover the most likely operational areas. The so-called 'Kammhuber Line' was established clear of the flak-defended areas, Germany's major cities being left in the care of the flak units. Boxes were positioned just inland, as roughly oval areas. New nightfighter crews were given a standard briefing from the 'old hands':

> The only possibility then is the trapping of the enemy in the searchlight beams. For this purpose in front of the Ruhr we have created a barrage of nightfighter sectors which are covered by two waves. Each of these sectors is provided with sufficient searchlight batteries to catch the enemy bombers in their beams when the weather is fine. Previously we had to circle at combat height over the beacon and try to spot the bomber flying at approximately the same altitude as ourselves. When one of them is caught in the searchlights we attack at once. We have to be careful not to overshoot the mark. The bombers fly at 220mph and we dive on them at between 280 and 310mph. Moreover, the Britisher is a sportsman but very tough. He sells his skin dearly, and as soon as he sees a night fighter he blazes away with everything he's got . . . You must always think: with each bomber, death and destruction fly over our cities. Protect your country, your women and children from death out of the skies. Put all your efforts into the defence of your country. [Wilhelm Johnen, *Duel Under the Stars*, Crecy, 1994, p.15]

The advent of ground radar, Würzburg A being introduced in the autumn of 1940, greatly improved matters, as a system basically similar to that perfected by the RAF was introduced. An experimental radar station was in operation at Nunspeet, Holland, by early September, working primarily with II/NJG1. The initial application of radar was in conjunction with the searchlights in the 'Helle Nachtjagd' system, whereby one Würzburg was used to control the nightfighter and another controlled a searchlight cone for illuminating the bombers. This system included a plotting room able to control three nightfighters, one for each zone. Similar zones were later established near Kiel and Bremen. Although the system achieved a modicum of success, the Germans found, as had the RAF, that such techniques were severely limited by bad weather and cloud. Radar had proved its effectiveness, so Kammhuber pressed for more equipment. The availability of radars and the introduction of new types, such as the Giant Würzburg, allowed a refinement of the basic box system to take advantage of GCI.

The 'Himmelbett' (Four-Poster Bed) system for use by 'Dunaja' (Dark Night) fighters consisted of a series of hemispherical areas, each some 50

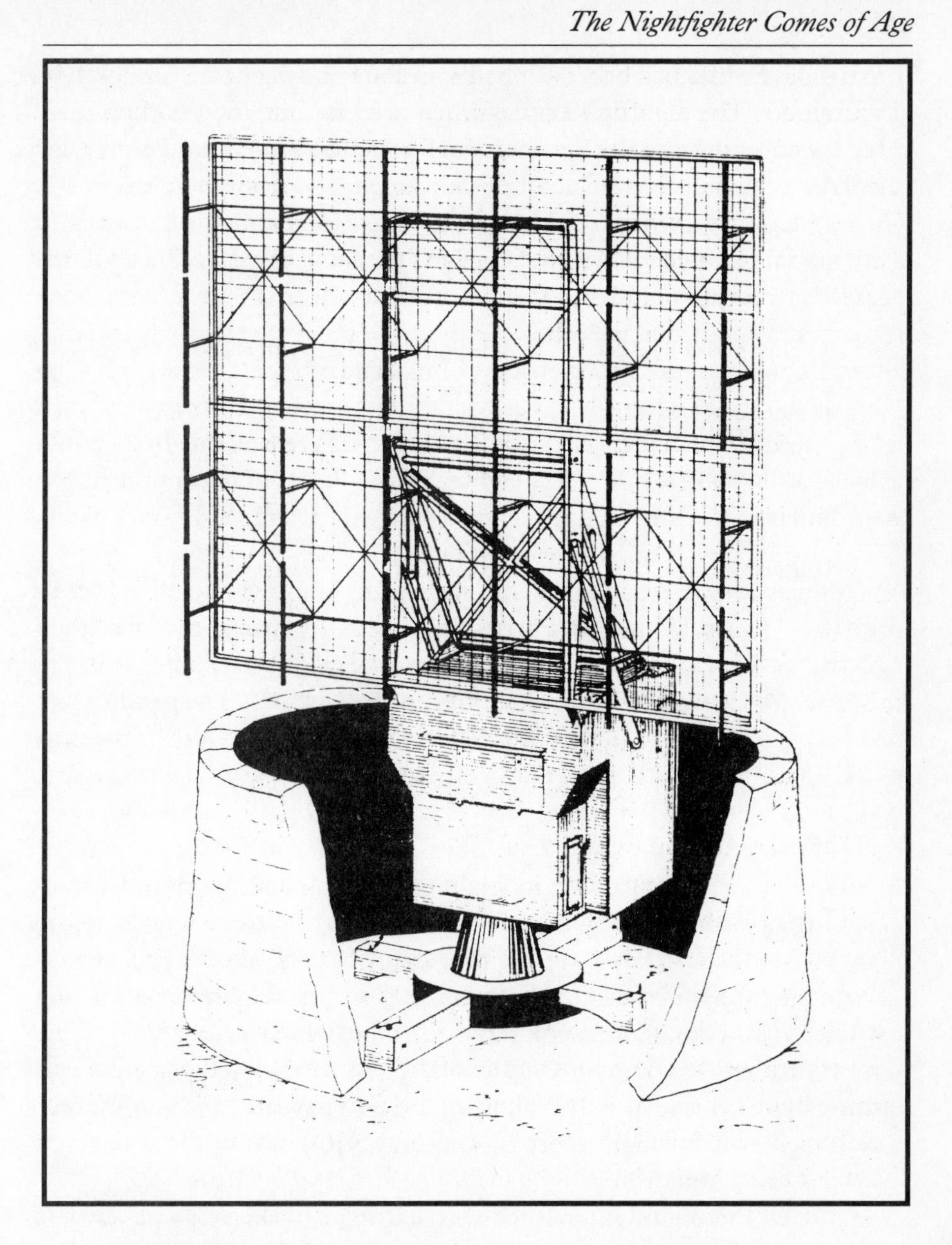

German Freya (FuMG 80) early-warning radar antenna

x 70km, equipped with two Würzburg radars and one Freya. The Freya (FuMG 80) had an approximate range of 62.5 miles for a target at 10,000ft (and a notional maximum range of 125 miles), and so was used for long-range search. In conjunction with information from other radar units, especially the coastal radars, it fed initial information into the control room. The FuMG 62 Würzburg had a notional range of 18.25 miles in its search mode, reducing to 11 miles when used for direction finding. Like Freya, it was produced in a number of sub variants. The control rooms, like the RAF's SOCs, were the heart of the system, and from them the FCOs (Fighter Control Officers) carried out their GCIs, talking the fighters into a position from which they could engage the enemy aircraft. The GCI was achieved by means of the two Würzburgs. One, the 'red', was used to track the enemy aircraft, and the other, the 'blue', to track the nightfighter. Each aircraft's position was displayed as a point of light on the 'Seeburg' plotting table. It was the FCO's task to work out the appropriate geometry and pass instructions to the fighter via a VHF radio link. A reasonable FCO could achieve a set-up that placed the fighter within 300m of his opponent (all very similar to British practice, as one would expect). On most dark nights over Europe, however, this was not enough.

This new system was the result, once again, of individual initiative. In mid-1940 Ernst Udet had devised a system for using two modified Würzburgs as a means of achieving ground control of the fighter, with, in this case, Falck acting in the role of the enemy bomber. The results were promising and the system was adopted as described above, although it was to be October before the 'Himmelbett' system could claim its first confirmed victory, Oberleutnant Becker of 4/NJG1 using his Do 17 to shoot down a Wellington on 2 October.

As the Luftwaffe expanded its night defences, a number of units were given the new role. Among the pilots of I/ZG76 ordered to covert to the night role in August 1940 was Helmut Lent. Like most of his colleagues he was far from enthusiastic, and over the ensuing months made many requests to his commanders for a transfer, all of which were refused. This was unfortunate for Bomber Command, for, as we shall see, this pilot was eventually to achieve over 100 night victories. That same month, Werner Streib increased his night score to four and NJG1 was in the process of adding a third squadron.

Although the nightfighter force was starting to achieve some success over the Continent, many considered that richer pickings could be had by taking the war to the enemy at his airfields. The German night intruder offensive started in the summer, and the fighters claimed a Hurricane on the night of 17 August (a kill that does not appear to be confirmed by British records). However, the first loss was suffered the following night when the 4/NJG1 Ju 88 of Feldwebel Schramm failed to return. More losses followed in August without any compensating successes, but the basic operational procedures were being established and the prospects still looked promising. A move to Gilze-Rijen in September brought a major reorganisation, the unit becoming 1/NJG2 but still operating the Ju 88C-2 (with a few Do 17Z-10s still on strength).

The standard procedure was to try and launch three waves of intruders, one to hit the airfields as the bombers took off, one to meet them over the North Sea, and one to be waiting when they returned to their airfields. The initial information on likely bomber activity would be provided by the intelligence services, often from listening to British radio frequencies. The southern part of England was divided into three operational areas: Raum A (Yorkshire), Raum B (Midlands/Lincolnshire), and Raum C (East Anglia).

The scale of operations remained small, and it was not until 24 October that the first confirmed kill was made, although five had been claimed by mid October. Oberleutnant Kurt Hermann and Feldwebel Hans Hahn were operating in Raum A, and in the early part of their patrol had attacked and damaged a Blenheim and a Beaufort. Approaching Linton-on-Ouse, they spotted a Whitley on take-off and promptly pounced on the hapless bomber, shooting it down. On the previous night two aircraft, including that of the Gruppenkommandeur, Major Karl-Heinrich Heyse, had failed to return from a patrol, so this victory was looked upon as partly redressing the balance. A few days later the Ju 88s also claimed a No. 49 Squadron Hampden. Although the intruder operations continued for the rest of the year, the balance sheet was not promising; twelve aircraft lost against eighteen claimed (but only two confirmed by British records). It must be stressed, however, that such basic statistics take no account of the impact such missions had on the operational status of the RAF. Were bomber missions lost? What was the effect on bomber crew morale? What strains were put on the RAF nightfighter force?

From September 1940 onwards the Luftwaffe records of VI Abteilung (the Quartermaster General's Department of the Air Ministry) give the weekly establishment and availability of nightfighters. Hitherto they had been included as part of the overall fighter total. The first statistics are those for 14 September, showing an establishment of 116 aircraft but an availability of only 55. Although the totals rise through the year, peaking at an establishment of 186 in late November, the serviceability does not improve, remaining at about 60 per cent. The German nightfighter serviceability rate remains low, down to 40 per cent at times, throughout the war. A tabular summary is given in Appendix 2.

As many of the nightfighter crews soon discovered, the British bombers were not easy targets. Gunners were always alert, and the fire from the turrets, especially the four-gun rear turrets, was often devastating. On the night of 15/16 October 1940, Bomber Command launched 134 sorties against various targets in Germany (mainly oil installations) and the Channel ports. A number of combats were reported by the bomber crews. No bombers were lost, but a study of German records shows just how effective the gunners had been. Two nightfighters were shot down by the RAF's air gunners; Bf 110D A9+FK of NJG1 (with the loss of two crew) and Bf 110D 2A+BL of 4/NJG2. However, the highest loss rate on night operations remained that of accidents on take-off, such as the ground collision between two Bf 110s of I/NJG1 and II/NJG1 at Schiphol on 26 October, or, more usually, on landing. October also saw Streib continuing to make his mark as the leading nightfighter exponent, achieving three victories on one night.

For the remainder of the year most Bomber Command attacks followed the established pattern of sending small numbers of aircraft to a number of different targets on the same night. There were a few exceptions to this, as on 7/8 November, when 63 aircraft attacked the Krupp works at Essen, on 16/17 November, when 130 bombers went to Hamburg and, more significantly, the major attack on Mannheim on 16/17 December. Called Operation Abigail Rachel, this raid was ordered by the War Cabinet as retaliation for the German attacks on British cities, and was to be mounted against the centre of a German city. A total of 134 bombers, the largest force to date but somewhat short of the planned 200 aircraft, took part, and just over two-thirds claimed to have hit the target. Mannheim's defences were weak, and the nightfighter force played little part. Bomber Command lost two Hampdens and one Blenheim (plus four other aircraft that crashed in England), but damage to the target was light. The gloves were now definitely off.

Chapter 4

AIRBORNE RADAR PROVES ITS WORTH

By early 1941 the RAF's GCI network for night operations was beginning to come together. With both ground controllers and aircrew needing to become familiar with equipment and procedures, the squadrons undertook intensive training by day and night. Among the pioneers were Flight Lieutenant John Cunningham and his AI operator, Sergeant I. Rawnsley. Daytime training was particularly valuable, as the pilot was in visual contact with the target throughout the latter stages of the interception and could, therefore, get a feel for the problems of his radar operator, although it was essential to avoid the temptation to put the aircraft in the correct place regardless of the instructions being given.

Six GCI sites were operational in January (Avebury, Sopley, Durrington, Willesborough, Walding Field and Orby), each associated with a sector control. The standard GCI technique was the 'curve of pursuit'. Using a map on which target and fighter tracks were plotted, the GCI controller positioned the fighter some 6–8 miles behind his target and on a co-altitude parallel track. He then transferred to his screen to calculate and monitor the actual curve of pursuit so that the fighter ended up astern at 2 miles, at which point the intercept could be taken over by the crew: 'It is very important that, when handed over to AI, the fighter should be directly behind the bomber, on the same course, at the same height and with an overtaking speed of not more than 30–40 miles per hour.' The latter aspect was one of the hardest to achieve; too fast, and the fighter would inevitably overtake his target without getting into a firing position.

The success rate for Fighter Command began to increase by March, the fighters claiming 15 aircraft during the 12-14th of the month. The figures for the first six months of 1941 looked promising, with 100 German aircraft falling to GCI-controlled intercepts.

The German attack on Liverpool during the night of 12/13 March was typical of the offensive operations being undertaken at this period. Some 169 aircraft of Luftflotte 3 were scheduled to take part in this raid, although it appears that only 146 found the target, along with 170 aircraft from Luftflotte 2 (including, according to some records, a few bombers from Luftflotte 5). Those bombers that found Liverpool dropped 303 tons of bombs plus 1,982 containers of incendiaries (most being of the standard 36 x 1kg variety). Ministry of Home Security records state that there were 408 fires in the Birkenhead area and 126 in Liverpool. Although there was a full moon and airborne visibility was good, ground visibility was only moderate. Nevertheless, the German crews had little trouble finding the coastal target and the first bombers were over the target at 2040. This was the first of nine distinct waves of attack that lasted for some six hours. German records show that it was delivered as a multi-axis attack in order to split the defences, although routeing does not appear to have taken much account of the AA areas, and aircraft flew over defended locations en route to Liverpool. Many German crews reported sighting fighters and it was, indeed, a busy night for Fighter Command, with 178 sorties flown by the six Groups – 32 by 9 Group, 19 by 10 Group, 59 by 11 Group, 48 by 12 Group, 15 by 13 Group, and 5 by 14 Group. The following claims were recorded in the Fighter Command log:

9 Group:			
96 Sqn	Hurricane	He 111 destroyed	Crashed at Wychbold
307 Sqn	Defiant	He 111 damaged	South Coast
10 Group:			
604 Sqn	Beaufighter	Ju 88 destroyed	Crashed nr Warminster
604 Sqn	Beaufighter	?Probable	Over sea
604 Sqn	Beaufighter	Ju 88 damaged	South Coast
604 Sqn	Beaufighter	He 111 damaged	South Coast
11 Group:			
264 Sqn	Defiant	He 111 destroyed	Crashed in sea
264 Sqn	Defiant	He 111 destroyed	Crashed Beachy Head
219 Sqn	Beaufighter	Ju 88 probable	Over sea
12 Group:			
255 Sqn	Defiant	He 111 probable	Retford
151 Sqn	Hurricane	Ju 88 probable	Over sea

The No. 96 Squadron ORB recorded the night's activities:

18 trips hunting for the enemy and the result – no large numbers of enemy aircraft blazing on the ground, but just a drawing of enemy blood in 'probables', and a squadron with tails well up and a few gunsights and gun muzzles that had spat forth fire at the enemy machines. There was great enemy activity over Liverpool and several of our aircraft were in action for the first time. F/O Vesely was the first in action, having taken off in Defiant N1803 at 2155 with Sgt Heycock to patrol Cotton East at 15,000ft. He saw

an He 111 above on the port side and told the air gunner, but the guns failed to fire. He kept the Defiant in formation with the German aircraft and flew alongside and slightly below expecting that the air gunner would get the guns to fire. Then the pilot of the Heinkel dived, followed by F/O Vesely, who manoeuvred to get on to the starboard side. He flew in formation again but the side gunner of the German bomber got in two bursts. Pilot felt that he had been hit in the chest, shoulders and left arm. He lost consciousness and when he came to found the Defiant falling in a spin; however, he managed to recover and return to land despite his injuries.

The No. 96 Squadron victory went to Sergeant McNair in Hurricane V7752. Another of the squadron's Defiants put in a claim for a probable, Sergeant Taylor and Sergeant Broughton having chased an He 111 into the Welsh mountains at low level after their guns had jammed (after firing only six rounds and despite the best efforts of Broughton to clear them).

German records of Luftflotte 3 show six losses (an He 111 of II/KG27, two He 111s of KG55, two Ju 88s of II/KG76, and an He 111 of III/KG26), plus two others that crashed on the way home. There are no records for Luftflotte 2. The AA guns of Merseyside and Birmingham laid claim to four aircraft. Other German aircraft were active, attacking various airfields, although only Tangmere, Coltishall and St Eval recorded any enemy activity. Because of the incomplete nature of the German records it is not possible to sort out the fighter and gun claims, but it is true to say that the nightfighters, both AI-equipped and 'cat's-eye', had performed well. If it is assumed that losses in Luftflotte 2 were of a similar magnitude, as a similar number of aircraft were involved, then the German losses were in the order of 18 aircraft.

AI-equipped fighters had flown 38 of the defensive sorties and claimed one destroyed, two probables and two damaged, a very respectable tally. However, an ORS report of 18 April really shows the advantage of AI-equipped aircraft over their single-seat colleagues. This report covered the period 2/3 March to 12/13 April, when the twin-engined aircraft (Beaufighter, Blenheim and Havoc) flew 492 sorties with a total claim of 98 contacts, 64 combats and 33 enemy aircraft destroyed. The single-engined aircraft (Defiant, Hurricane and Spitfire) flew 1,211 sorties resulting in 78 contacts, 54 combats and a claim for 24 aircraft destroyed. The report states:

> Owing to the lack of certain returns it has not been possible to assess the relative efficiency of different methods of nightfighter control; but it may be reasonably assumed that the best control at present in use is that which employs GCI in conjunction with the AI-equipped Beaufighter. Of the

> average results obtained by this combination, under conditions of bright moonlight and clear visibility during the months of March and April, it may be stated that of the attempts to intercept enemy aircraft, 15 per cent resulted in combats, and of these combats 70 per cent resulted in the destruction or probable destruction of the enemy aircraft.

These results held high promise, and better radars were in the pipeline.

Cat's-eye fighters were still achieving success. In the early hours of 8 April, Squadron Leader J. Simpson, OC No. 245 Squadron, was airborne in a Hurricane from Aldergrove with orders to patrol Ardglass Donaghadd:

> I was in the Mess when news came that the Germans were dropping bombs on a town nearby. It seemed rather strange. The war had not come so close to Ulster before. Incendiaries had been dropped and high explosive bombs were on the way. I was next to patrol. It was about 1.15 in the morning, dark with a sickly moon shining through a mist. I took off and climbed to about 9,000ft, passing above the clouds into another world, where the moon, in its second quarter, shone out of a blue-black sky.
>
> I was told that there were aircraft near me. My eyes searched the blackness. There was no horizon: no object upon which to fix one's eyes. And one had the illusion, travelling at 200 miles per hour, that every one of those brilliant stars was the tail light of an aircraft. I searched among that moving pattern of lights and my eyes rested upon two black objects. I could see them because, as they moved, they obliterated the stars. They were quite near when I recognised them as aircraft . . . whether enemy or not, I was unable to tell. So I flew nearer and learned soon enough. The rear gunners of both aircraft fired a shower of bullets at me, some with whitish-green light of tracer bullets, some glowing red. They missed me and for a minute I lost them. Then I saw them again, farther apart, moving against the white floor of the clouds below me. They were black and quite clear. The advantage was now mine for they were perfectly placed as targets. I crept down to attack the rearmost of them. They were flying slowly. It was difficult for me to withhold my speed so that I would not overtake him. At a distance of about 200 yards I opened fire from slightly below. Then came my next surprise . . . the blinding flash of my guns, in the darkness. In day time one does not see it. At night it is terrific and I was so blinded that I lost sight of my enemy. I broke away and lost him for a few seconds. I next saw him going into a gentle dive towards the clouds. The increase in speed made it easier for me to attack and I closed in to 80 yards. I opened fire once more. This time I was prepared for the flash and kept my eyes on the enemy. His rear gunner returned my fire, but only for a second. I had apparently got him for he was silent after that. I continued my fire, closing in to about 50 yards. Then I saw a comforting red glow in his belly. I was still firing when the Heinkel blew up, with a terrific explosion which blew me upwards and sideways. When I righted myself, I was delighted to see showers of flaming pieces . . . like confetti on fire . . . falling towards the sea. I was able to enjoy the satisfaction of knowing that

> I had brought him down before he had released his bombs. The second Heinkel had disappeared and I asked for homing instructions over the wireless.

Simpson was duly credited with an He 111, and another swastika joined the other ten on his Hurricane, either side of the 'Joker' cartoon.

Another night success, this time a Ju 88, was achieved on 6 May:

> Sqn Ldr Simpson took off from Aldergrove at 0020 hours and was detailed to patrol Ardglass. Whilst circuiting over position he was vectored by controller to a raid approaching from the south. Sqn Ldr Simpson saw 3 bandits in a straggling line astern formation immediately ahead of him. He thought the E/A were flying south and he at once opened up on the nearest. He then realised that the E/A were flying straight towards him and that he was making a head-on attack. He continued to fire and saw the E/A's nose and engines burst into flames as it dived into the sea. Sqn Ldr Simpson dropped a flare hoping to use his cine gun but was immediately attacked by the other two E/As.

These two accounts well illustrate the range of problems faced by the cat's-eye nightfighters in their attempts to locate and destroy the enemy.

Although the AI-equipped squadrons were gaining experience and the system showed great potential, there were still many problems to overcome. A major one with the AI phase was that of 'weave', caused by a combination of blip accuracy on the scope and the time delay between a course correction being requested and the aircraft actually taking it up. Obviously, the closer the fighter got to its target the more exaggerated this became. Although this could be offset by experience, the pilot adopting a mean course, it was considered that the provision of a pilot's indicator displaying the target blip would enable him to damp out the weaving by keeping the dot roughly central. The earliest examples went into use in July, but were soon replaced by a revised version incorporating a range indicator. With AI Mk. IVA the radar operator would select the target blip, track it and then feed it to the pilot's 'G Scope'. The practical application of the system came at the very final stages, when any 'weave' might otherwise result in a failed intercept.

As with so much of this early equipment, this was largely a hand-built system, and thus it suffered from such problems as inadequate manufacturing tolerance. The Mk. V system, tested in Havocs and Beaufighters from May 1941, was, in essence, a factory-built version that worked well but had a high unserviceability rate. It was far too complicated a set, requiring higher maintenance and technical support than was available at most front-line stations. (The first AI Mk. V-equipped Mosquito squadron was No 157, in April 1942.) Meanwhile, Blenheim N3522 was being used to flight-test the prototype AI (known as AIS) using the cavity magnetron developed by GEC and the TRE. This would later enter service as the AI Mk. VIII. According to Air Ministry Air Publication AP1093D:

> AI Mk. V was designed so that the operator need not instruct the pilot orally during the interception. The pilot is provided with an indicating tube of his own, and on this the information is displayed in a manner which can be easily read. In operating AI Mk. IV the observer must interpret his picture and tell the pilot what to do. This introduces a time delay which must be cut to a minimum if a dodging target is to be successfully followed. There are several sources of error; the operator may misinterpret the picture; he may give the wrong instruction; the pilot may fail to hear the observer's message, and a repeat may be necessary. In AI Mk. V an operator is still required to work the equipment but the information is relayed electrically to the pilot. The desirability of a pilot's indicator is a highly controversial question. There are a few disadvantages, for example, the pilot must divide his attention between looking out into the darkness for the enemy and watching the tube. Up to the present, the method of giving oral instructions, and leaving the pilot free to look for the enemy, seems to be the most successful.

While the nightfighter's performance was only marginally better, if at all, than that of its targets, the speed and accuracy of the intercept would continue to play a major part in ensuring success; any small error and the bomber would almost certainly escape.

A number of solutions were tried to overcome the shortcomings of the existing CH and CHL stations that were employed for GCI work. The most promising trial was that held by the Kenley Sector in April 1941, with the use of gun-laying sets. While such sets were accurate and gave a height indication, they were also short-range, so an area cover required the use of a 'carpet' of sets, with consequent problems of integration and co-ordination. During April the Defiants of No. 264 Squadron flew 76 such sorties resulting in 56 attempted intercepts, but only seven visuals, with claims for four enemy aircraft destroyed. It was concluded that the concept held promise, but that the limited capacity and the difficulties of co-ordination would restrict its operational employment.

This saturation of the ground radar and control network was a problem for all nightfighters, and the only solution was to let the fighters roam free, using either freelance AI contacts or cat's-eye techniques. The freelance AI system was seen as:

> . . . the most effective method of employing AI-equipped fighters under conditions of heavy raid density. The establishment of a series of transverse patrols at stages along the route to the target in order that those raiders that were pursued by one patrol but escaped destruction, would be successively attacked by the remaining patrols.

It was, however, a very hit-and-miss affair, and produced little result in return for the effort expended.

The stage is now set, with four basic styles of nightfighter employment: single-engined fighters on GCI and cat's-eye; twin-engined fighters with AI on GCI or freelance. These techniques, modified to suit conditions or equipment, will recur throughout the nightfighter story.

The Duxford Wing was determined to prove that the Spitfire was suitable as a nightfighter, so No. 266 Squadron undertook a number of missions. Overall, the results were poor, the Spitfire being very prone to problems on landing at night. However, a number of successes were recorded during May.

The overall scale and success of Fighter Command's night operations continued to increase, as borne out by the figures given in Table 2. The Air Ministry narrative concluded:

> While the Fighter Night might not have been an ideal method of night interception, it brought results at a time when they were badly needed, and so long as the conditions of visibility were good, it was decided to continue with them whenever opportunity presented. . . In night defence the importance of precise control and the importance of specialisation had been well demonstrated, both these things becoming possible by the application of scientific principles to the problems we faced.

The AI nightfighters were achieving far more contacts for sorties flown (over 30 per cent, as against less than 10 per cent for the cat's eye), but only a third were converted into combats. Aircraft performance was still a limiting factor.

The foregoing account has concentrated primarily on the countering of the Germans' bomber offensive, but their intruder campaign was also still in full swing. Although German losses remained high, there were an increasing number of successful nights, such as the early hours of 11 February 1942, when the intruders of 1/NJG2 were waiting for the RAF bombers over their bases in England. Ten of the unit's nightfighters were on patrol when the bomber force, primarily Wellingtons and Hampdens that had been bombing Hannover, returned. The bomber force had already lost four of its number over Europe. In a short time the intruders

TABLE 2. FIGHTER COMMAND NIGHT OPERATIONS, JANUARY–MAY 1941

	AI fighters			Cat's Eye fighters		
	Ops	Contacts	Combats	Ops	Visual	Combats
January	84	44	2	402	34	9
February	147	25	4	421	33	9
March	270	95	21	735	34	25
April	542	117	50	842	45	39
May	643	204	74	1,345	154	116

claimed to have shot down ten bombers, although RAF records show only three that can be attributed to these attacks. Nevertheless, the Germans returned home without loss, having caused confusion and no small amount of panic. It had been a good night, and one that showed the promise of the technique if only more fighters were available.

The confusion and fear among the bomber crews, caused by the presence of the nightfighters, is difficult to quantify. Most bomber crews recall that, after the strain of hours of flight, having faced flak and fighters over Europe, they considered themselves safe once they crossed the English coast. To most, the biggest danger as they neared their airfields seemed to be the risk of mid-air collisions with other bombers; hence the use of navigation lights, even after ground control had passed the 'rats' warning that enemy nightfighters were present.

On the nights that the intruders were active they were often more likely to come into contact with RAF training aircraft. Such was the case on 26 February, when an Airspeed Oxford was shot down by Oberleutnant Hermann in his Ju 88, R4+CH. However, he in turn was shot down by flak, crash-landing in the King's Lynn area in the early hours of 27 February. As was usual when the aircraft was in reasonable condition, the Ju 88 provided useful data for the RAF's intelligence section; in this case its additional MG151 installation proved to be of interest.

The intruders continued to operate at the rate of about twelve sorties each month throughout the spring and summer. Losses once more began to rise as successes diminished, although the unit laid claim to its 100th victim during this period. This was considered to be the result of increased RAF nightfighter activity, which, as outlined above, was true. However, it was also due to what could be called 'operating hazards'. The first of these was the danger from the debris of the enemy aircraft. It was a natural instinct at night to close to very short range and make sure of the kill in what might prove to be the only available moment; thus the danger of self-damage was higher. The other danger was that of mid-air collision, either through an error of judgement when closing for a kill or through sheer accident. One of 1/NJG2's leading crews was lost in this way when the Ju 88 of Leutnant Volker collided with Wellington R1334 of 11 OTU near Ashwell, Hertfordshire. Then, on 12 October, Kammhuber visited the unit at Gilze-Rijen, not only to praise them for their efforts, but also to tell them that an immediate move was ordered, to Sicily for operations in the Mediterranean theatre. This phase of the intruder campaign was over.

During the early part of 1941, Bomber Command aircraft continued attacks on oil installations such as those at Gelsenkirchen and Homberg, plus other industrial targets and ports. By March the three new heavy bombers had entered service, the Short Stirling, Avro Manchester and Handley Page Halifax, virtually doubling bomb loads and greatly increasing the RAF's offensive capability. Attention was also paid to self-defence, and the four-gun rear turret was a standard feature, along with nose and mid-upper turrets, although the .303 machine-gun remained the standard weapon.

From March to July the major offensive effort was against naval targets, the warships *Gneisenau* and *Scharnhorst* at Brest being particular favourites. Most losses were due to flak, although nightfighters did venture to the western areas on a few occasions. Some of the day fighter units were also called on to mount night patrols. Naval-related targets, especially U-boat factories and depots, in Kiel, Hamburg and other cities were also high on the list, and although nightfighters were more active in these areas, the highest percentage of bomber losses was still to flak. As routeing avoided the highly defended Ruhr area, overall loss rates were kept low. A typical mission was that to Hamburg on 10/11 May, when 119 bombers took part. The weather was good and the bombing reasonably concentrated; only four aircraft, three Wellingtons and one Whitley, were lost. Overall, the greater number of bomber losses were due to crashes in England owing to shortage of fuel, bad weather, battle damage, or a number of other causes.

Although Luftwaffe nightfighters achieved a number of successes, their loss rates continued to be high, British nightfighters becoming an increasing problem for the intruder missions. Typical of these losses was that of Ju 88 R4+DM of 4/NJG2, flown by Unteroffizier Richard Hoffman and crew. The aircraft took off from Gilze-Rijen at 2315 on 13 June, armed with eight 50kg bombs. Arriving over East Anglia, the crew began to search for targets, but at 0020 the aircraft was suddenly hit by a heavy burst of fire from behind which caused it to catch fire straight away. The Ju 88 was obviously doomed and the pilot and bombardier baled out, but the wireless operator did not have a parachute and was killed in the crash. The intruder fell to earth at Narford, near King's Lynn.

It was the same scenario a week later, when another Ju 88, R4+1H, was also destroyed by a nightfighter whose presence the German crew did not suspect until a devastating fire was opened. This aircraft crashed at Haines farm, Deeping St James. Yet another, R4+LK, flew into a hillside at Barnby Moor, Yorkshire, in the early hours of 4 June, with the loss of all three crew. This last crash provided a bonus to the British intelligence services in the form of a log book which not only showed that one crewman (Gerhard Denzin) had flown 30 operational sorties since the previous November but, more interestingly, gave details of his training. Initial training, undertaken at Nordhausen, comprised 27 W/T training flights in Ju 86s. It was followed by a few days of air gunnery, using the Fw 58, at Malacky, Slovakia. The OTU course was also short, less than a month, and included bombing (Ju 86 and He 111), navigation (Do 23), low level (Do 17) and air combat (Fw 58), but with a total flying time of only 13hr. The next stage of training was night flying, which took place at Brandir, followed by posting to the operational unit at Gilze-Rijen.

Despite pressure from some quarters to increase the level of intruder activity, many senior voices were raised against the proposal, some on the grounds that public morale was better served by seeing 'terror' bombers falling to earth over Germany, and others because of the rising loss rate, and uncertain success, of the intruders. Generaloberst Hans Jeschonnek, Chief of the Luftwaffe General Staff, did not support the policy, and on the direct orders of Hitler the intruder campaign came to an end in October.

Meanwhile, the only other option, and one that was already under way, was to expand and improve the home defence network. On 1 August the nightfighter organisation was raised from Division to Korps status (as Fliegerkorps XII) under the leadership of Kammhuber, with its HQ at Zeist, near Utrecht.

The nightfighter squadrons worked hard to improve their chances of success:

> The pilots, both officers and NCOs, usually swapped experiences with their comrades in blacked-out waiting rooms. Models of enemy aircraft were often projected on to the ceiling with the aid of a bright torch in order to familiarise the crews with the enemy types. This lightning recognition of different opponents was decisive for the first attack, since each enemy machine had

> its own strengths and weaknesses and, above all, a different defence armament. The most dangerous was the rear gunner, who sat with four heavy machine-guns in a Perspex compartment behind the tail unit, waiting for the attacking nightfighter. These 'tail-end Charlies' were usually chosen for their courage, for the job made the greatest demands on the nerves. The man sat for six to eight hours entirely alone in his narrow unarmoured cage, and had to be on the alert the whole time lest a German nightfighter should suddenly bob up out of the darkness and try to destroy him first with a well-aimed burst. If he were hit the fate of his machine was usually sealed. [Johnen, *ibid.*, Chapter 4]

A significant development that same month was the introduction of the AI-equipped Bf 110, which carried the Lichtenstein B/C set. The first production sets were tested at Rechlin, and although the test pilots found few problems with the AI itself they did make adverse comments about the drag (and loss of speed) caused by the aerial array. The trials then moved to II/NJG1 at Leeuwarden, where at first the operational crews were very sceptical.

Flying on the night of 9 August, Leutnant Becker and Sergeant Staub picked up a contact on the equipment at 2,000 yards. Staub talked his pilot towards the target, but it kept slipping outside the 25-degree coverage of the radar. This happened a number of times, and each time Staub had to guess which direction to turn in order to reacquire the target, until eventually they shot it down. By 30 September he and his crew had shot down six bombers using AI. The Bf 110 was the Germans' primary AI-equipped nightfighter for much of the war and it remained reasonable effective. Although its performance limitations led to pressure to acquire greater numbers of Ju 88s, the latter type was in great demand elsewhere as a dive bomber.

After a break of four months (the anti-naval campaign), Bomber Command resumed its Strategic Bombing Offensive with a directive to disrupt the transportation system and destroy the 'morale of the civilian population and of the industrial workers in particular'. The weight of attacks steadily increased, and as a consequence the defences received greater impetus to achieve success. German air resources were being stretched, but the Nazi leadership had realised the importance of putting an end to the nightly visits by Bomber Command. Cologne, Hamburg, Mannheim, Essen, Hannover, Frankfurt, Nuremberg, Berlin and other German cities were on Bomber Command's list that autumn, operations taking place on most nights. The scale of effort gradually increased, almost 400 aircraft being involved on the night of 7/8 November; 169 went to

TABLE 3. INTRUDER – SUMMARY OF SORTIES AND CLAIMS

	Nts	Ble	Hav	Bos	Hur	Def	Tot	e/a obs	e/a des	a/f att	Cas
January	7	23	–	–	–	–	23	11	–	13	1
February	6	8	–	–	–	–	8	19	–	3	–
March	11	49	–	–	2	–	51	51	3	28	2
April	16	33	16	–	2	6	57	33	2	17	1
May	13	1	56	–	11	9	77	128	11	38	–
June	14	–	48	–	6	–	54	4	1	19	–
July	18	–	87	–	8	–	95	14	–	56	–
August	17	–	61	–	4	–	65	27	–	37	2
September	10	–	35	–	4	–	39	13	2	17	1
October	12	–	28	–	7	–	35	13	1	23	–
November	8	–	21	2	4	–	27	9	–	11	–
December	10	–	22	1	5	–	28	32	1	15	1
Totals	**142**	**114**	**374**	**3**	**53**	**15**	**559**	**354**	**21**	**277**	**8**

Nts = nights; Ble = Blenheim; Hav = Havoc; Bos = Boston; Hur = Hurricane; Def = Defiant; Tot = Total of all types; e/a obs = enemy aircraft seen; e/a des = enemy aircraft destroyed; a/f att = enemy airfields bombed; Cas = own aircraft losses.

Berlin, and 21 were lost. This loss rate of 12.4 per cent was more than Bomber Command could sustain, it being generally accepted that a sustainable loss rate was less than 5 per cent. Berlin, however, was a particularly hard target. Not only were the flak defences of the city itself strong, but the bombers had a long flight over enemy territory, increasing the chances of interception by nightfighters.

Throughout the latter half of 1941 Kammhuber continued to refine his basic system in the west, extending the coverage of his 'boxes' and providing purpose-built operations rooms (known as 'Kammhuber's cinemas'). However, even with such refinements the system remained limited by its ability to handle only one target (and fighter) per box. It was also found that the boxes were not deep enough: the enemy bombers crossed the danger zone too quickly. As the RAF's bomber operations continued to grow in scale, the defence system was rapidly swamped. Nevertheless, RAF planners were growing concerned over the dangers.

During 1941 Bomber Command had operated on 240 nights, dropping 31,700 tons of bombs. Yet it had been a frustrating year in which proof had been provided that only a small percentage of bombs were falling anywhere near their targets. This, added to the rising losses, led many to question the future of the bombing offensive. It was essential for Bomber Command to improve results and reduce its losses. It thus became important to disrupt or destroy the enemy night defences, and it was this desire that led to the expansion of the RAF's night intruder campaign – with the dual intention of attacking enemy night bomber airfields.

Although there was still only one squadron, No. 23, fully engaged on night intruder operations (this unit flew 488 of the 559 intruder sorties recorded for the 142 operational nights of 1941), the concept had certainly become well established by the spring of 1941. The unit had re-equipped with Douglas Havoc Is in March 1941 to increase the offensive power of its operations, the first sortie with its new aircraft being flown the following month. In a postwar study, the Air Historical Branch summarised this period of intruder operations:

> Sighting of aircraft, known or believed to be hostile, was reported on 60 nights and altogether the number reported seen in the air or on the ground amounted to about 360. No. 23 Squadron claimed the destruction of 14 and the 'irregular' squadrons of Hurricanes and Defiants of seven. Nearly 300 bombing attacks on airfields were also recorded. There is evidence that such operations frequently caused the enemy to divert his returning aircraft to airfields other than their bases and that his accident rate increased in consequence. No. 23 Squadron lost eight aircraft on operations, six of them in the first four months of the year.
>
> The rate of destruction of enemy aircraft was therefore high in comparison with that achieved in other fighter operations and in proportion to the number of sorties flown. Throughout 1941 the planning and execution of intruder operations was carried out by various groups. A more centralised arrangement was advocated on several occasions but at the end of the year the problem remained unsolved.

The first operation had been flown on the night of 2/3 January 1941 by six aircraft, although three of these returned early with technical problems. One claim for a 'probable' was made. However, it was not until 3/4 March that the first claim of an enemy aircraft destroyed was made. On the same night one of the Blenheims failed to return. The 'score' for the year to that date was one enemy aircraft for three Blenheims which failed to return. The rate of claims increased through the summer months (see Table 3), and the intruder role was expanded with the involvement of the Hurricanes of No. 87 Squadron, whose first operation was flown on the night of 14/15 March, and No. 141 Squadron, whose Defiants first flew offensive patrols on 7/8 April. The first mission by No. 87 Squadron was a success, two aircraft visiting Caen/Carpiquet, where they found twenty aircraft on the ground, though in the prevailing conditions they claimed only one.

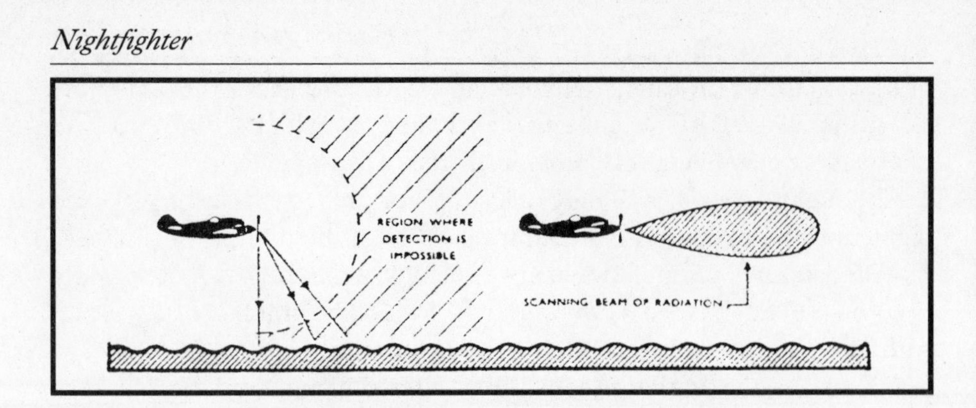

AI radar: ordinary (left) and 9cm-wavelength (right)

By the end of the summer a number of other units had also become involved: No. 264 Squadron (Defiants; May), No. 601 Squadron (Hurricanes; May), No. 242 Squadron (Hurricanes; June), No. 306 Squadron (Hurricanes; June), No. 3 Squadron (Hurricanes; July, August and October), No. 247 Squadron (Hurricanes; September, October and November), No. 253 Squadron (December), and No. 151 Squadron (December). Typical of the intruder operations flown by No. 23 Squadron was that of 13 September:

> S/L Hoare took off for Melun but, seeing lights near Beauvais, turned aside to investigate and found the airfield lit up. He first chased an aircraft with navigation lights that was landing but gave it up in favour of an aircraft seen orbiting, and the air gunner put in a long burst at 75 yards' range from underneath and to beam, but without visible result and the aircraft was then lost. Then another was seen and formated on and the air gunner put in a long burst until blinded by a flash from his own gun. The E/A put out its lights and was lost. It was claimed as damaged. Yet a third aircraft was seen and followed as it turned in to land. The pilot opened fire at 100 yards causing both engines to smoke strongly as the E/A fell away to starboard in a vertical dive at 600ft. It is claimed as destroyed.

The main motivation behind these missions was the same as that of the Luftwaffe intruder operations: to attack the bombers during a critical phase of flight when their awareness might also be lowered. However, for the RAF the impetus for such missions took on a much more intense nature as part of the campaign against the increasingly effective Luftwaffe nightfighter force.

The most significant technological development in 1941 was the adoption of centimetric AI sets. Although the theory of employment of such wavelengths had been proposed as early as 1936, it was not until the development of the Randall-Boot cavity magnetron valve, giving 9.1cm/3,300 megahertz, that it became a practical proposition. Improvements were made at the General Electric Company laboratories and the first trials were flown, by Blenheims, in March 1941. A parabolic mirror was used with a 12-degree beam, giving a spiral scan over a 90-degree look angle, the net result being much greater discrimination and accuracy. In combination with a single CRT presentation, the system looked very promising, and by mid-1941 trials with the Beaufighter had been ordered. As AI Mk. VII (initially designated AIS Mk. I), the new equipment was ordered into production and began to filter down to the squadrons, although in this interim set certain items, such as IFF, were omitted in the haste to get the equipment into service. Only a limited number of Beaufighter squadrons received this AI, but they achieved some notable successes.

A further development, AI Mk. VIII, underwent trials late in 1941, with modifications to make it more suitable for engaging low-level targets. The first operational unit was No. 29 Squadron at West Malling. At about the same time the FIU had carried out trials on the American-developed SCR.520, flying it in a Canadian Boeing 247D. Orders were placed, under

BRISTOL BEAUFIGHTER

The Bristol Beaufighter was to take on many roles during its wartime career, but none was more important than that of nightfighter. Although the Blenheim had proved the validity of airborne radar, it lacked the performance and firepower to be a truly effective nightfighter. Yet another private venture, the prototype Beaufighter, R2052, first flew in July 1939. Development was rapid, and the Fighter Interception Unit at Tangmere, Sussex, received its first aircraft in August the following year. Squadron service began the following month, and the first sorties were flown on 17/18 September. The first confirmed victory was by No. 604 Squadron on 19 November. Improvements followed, the Mk. IIF entering service in April 1941 with No. 600 Squadron, by that time one of the most experienced of AI units. This was in turn followed by introduction of the Beaufighter VIF. As AI radar variants were introduced, so the noses of the Beaufighter nightfighters changed, from the simple 'feathered arrow' aerials of the early sets to the 'thimble nose' that hid the later sets.

As the night war progressed, the squadrons were employed in an intruder role to hunt the enemy, in the air and on the ground, over his own territory. A variety of new devices were introduced as the electronic war became ever more complex. While the 'Beau' proved to be a sturdy and reliable aircraft, it also lacked the manoeuvrability for its new 'hunter-killer' task and so, in due course, it gave way to the Mosquito.

Beaufighter VIF data

Powerplant: Two Bristol Hercules VI. Crew: Two. Length: 41ft 8in. Wingspan: 57ft 10in. Maximum speed: 333mph. Ceiling: 15,000ft. Armament: 4 x 20mm cannon, 6 x .303in machine-guns in wings.

the designation SCR.520(UK), as an insurance against the British microwave developments running into problems. In the event this did not happen, and this initial order was cancelled.

The continued shortage of AI sets and suitable aircraft meant that single-seat fighters remained an essential element of the night defences. A Fighter Command document dated 2 May 1941 reported:

> During the past month considerable success has been achieved by 'Fighter Nights' patrols when conditions of moon and weather have been favourable. Under such conditions, fighters have been found capable of inflicting heavier losses on the enemy than AA fire during the same period. On the other hand, in conditions of darkness and low visibility, and in particular when the moon is below the horizon, it has been found that fighter pilots operating in 'Fighter Nights' patrols seldom sight the enemy. In such circumstances it is clearly preferable to give freedom of action to the AA guns as, under these conditions, they have a better chance of bringing down enemy aircraft.

The document went on to issue new operational instructions for the employment of such fighter patrols. The basic concept was that 'Fighter Night' operations would only be ordered in suitable moon and weather conditions, and only when 'the enemy attack is obviously concentrated on a clearly defined target which covers a limited area'; and, furthermore, that such fighters would only patrol over the target and not be vectored to other areas. The single-engined fighters would be given patrol heights over the target area, with the view that 'at least 20 fighters should be employed simultaneously'. If the target was gun-defended, then the AA guns would be restricted to a burst height 2,000ft below the lowest fighter patrol layer.

Numerous techniques were developed in an effort to improve the chances of these aircraft making successful intercepts. Perhaps the most unusual was the use of an AI-equipped aircraft to act as an airborne illuminator. Great faith had been placed in the use of ground-based searchlights as aids to cat's-eye fighter operations, so, surely, a logical progression was to get the searchlight airborne. Thus was born the concept of the Turbinlite aircraft: take an AI-equipped aircraft, in this case the Havoc, and fit it with a searchlight. Working with a single-seat fighter or two, the formation would be positioned by GCI, the AI operator would then carry out the final line-up and at the critical moment the searchlight would be switched on. With the enemy aircraft bathed in light, the fighter, usually a Hurricane, would sweep in to make the kill. In theory it sounded simple and brilliant. Initial development work was carried out by No. 93 Squadron, although selected personnel were soon moved sideways to form

No. 1458 Turbinlite Flight. The remainder of the squadron, and aircraft, went on to form two further Flights.

On 22 May 1451 Special Havoc Flight was formed at Hunsdon around an ex-85 Squadron nucleus, to 'train four other Havoc Flights in the use of Helmore Havoc searchlight aircraft'. Training began straight away, using the Hurricanes of No. 3 Squadron, also resident at Hunsdon, as satellite fighters.

The purpose of the formation of 1452 Flight in July 1941, under Squadron Leader J. E. Marshall, was: 'to test the practical application and to fully develop the techniques of a searchlight aircraft fitted with AI operating at night, with one or two parasite fighters in formation'. The first two Havocs arrived on 2 July, soon to be followed by others, and on 10 August a detachment of No. 264 Squadron Defiants arrived to act as parasite (or satellite) fighters. The next few weeks were spent developing the basic techniques, and with the OC 1452 insisting that the system would never work really well until the parasite fighters were an integral part of the squadron. On 6 November the unit submitted a progress report in which a standard operational procedure was outlined:

> In this attack, the parasite, when given the word, dives forward so as to lose about 500ft. The AI operator keeps the Turbinlite pilot directed on to the target, and gives the pilot the word to illuminate when he can see by his tubes

Helmore Turbinlite

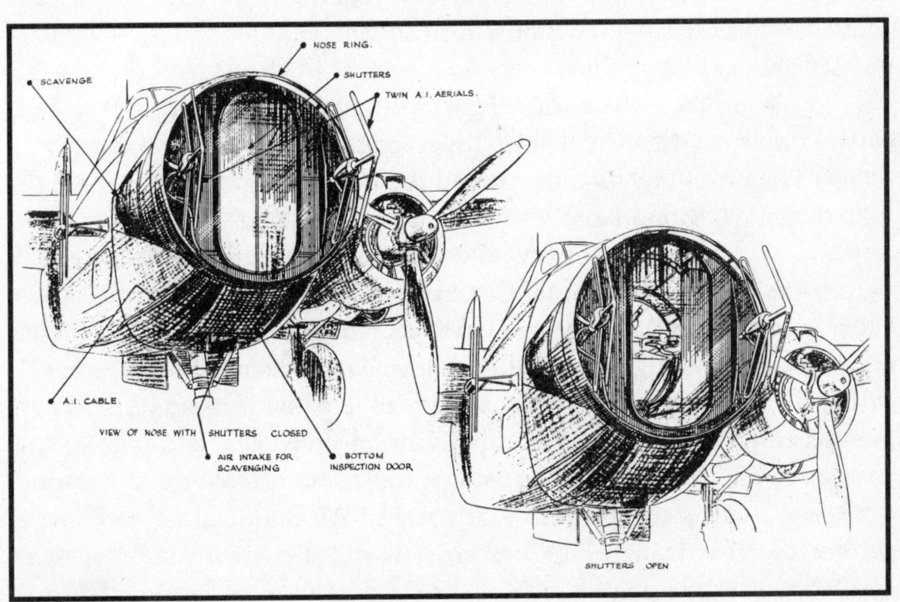

> that the parasite is approximately 300 yards behind the target. The parasite then sees the target illuminated slightly above him and is able to make a well-timed and effective attack.

So, at least, the theory said, and the unit had proved the basics of the procedure by day and night. Among the list of standard codewords, the three most important were 'Warm' (5,000ft range), 'Hot' (3,000ft range) and 'Boiling' (I am about to expose the searchlight in 10sec). A follow-up report stressed other important elements:

> Success in the scheme depends among other things, on each member of the crew taking the correct action at the appropriate moment. As orders are passed from one to another, and as the conditions of the intercept will seldom be identical on two nights, it is clear that a complete understanding must exist between each member of the team. The satellite pilot should always work with the same Turbinlite pilot . . . it is particularly easy in the first few seconds to miss an illuminated target. If the target is not illuminated, the beam should be raised about 10 degrees, and then depressed through 15 degrees.

By October 1941 there were five Turbinlite Flights and two more were in the process of forming. The last two months of the year were devoted to intensive training, and it was not until April 1942 that the system was cleared for operational employment.

In a refinement of the fighter/searchlight co-operation technique, a 'Fighter Box' system was instituted. The nightfighter would orbit overhead a central searchlight within a box (dimensions 44 miles by 14 miles), a series of boxes being joined together to form a belt. At the edges of the box – the Indicator Zone – searchlights would be positioned every 6 miles, whereas towards the centre – the Killer Zone – they would be every 3½ miles. When a enemy aircraft entered the box the nightfighter would be told of 'trade' inbound and given a height to put him 2,000ft above the target. The master searchlight would then indicate the direction of attack. On dark nights the codename 'Cracker' would give the fighter clearance to engage any aircraft illuminated by the searchlights. On bright moonlight nights the cat's-eye fighters would be allowed to freelance within the boxes without any searchlight control. This was to remain the basic system until the end of the war, with some variations and changes, although in the absence of any serious German raids it was never tested under pressure.

As can be seen from Table 4, the use of AI-equipped aircraft saw a marked increase from spring onwards, as did the effectiveness of their patrols. Up to this point much of the AI development and operational

TABLE 4. SUMMARY OF RAF NIGHT DEFENSIVE OPERATIONS, 1941

Date	Sorties		Contacts		Combats		Claims		
	German	RAF	AI	Vis	AI	Vis	NF	AAA	Other
January	2,295	84	44	–	–	2		12	2
		402	–	34	–	9	3		
February	1,820	147	25	–	4	–	2	8	5
		421	–	33	–	9	2		
March	4,125	270	95	20	21	10	15	17	4
		735	–	34	–	25	7		
April	5,125	342	117	10	50	5	28	39	2
		842	–	45	–	39	20		
May	4,625	643	204	13	96	6	37	31	10
		1,345	–	154	–	116	59		
June	1,980	536	94	–	37	–	20	12	3
		942	–	24	–	15	7		
July	1,352	557	80	–	23	–	20	1	6
		338	–	25	–	18	6		
August	935	549	82	1	4	–	3	3	4
		592	–	5	–	1	–		
September	838	361	74	3	10	–	7	1	2
		344	–	7	–	3	1		
October	849	621	106	10	33	3	9	1	2
		496	–	10	–	3	2		
November	695	417	80	4	14	1	7	3	1
		345	–	8	–	–	–		
December	695	440	46	1	6	–	3	3	3
		211	–	2	–	–	–		
		4,967				27	151		
		7,013				238	107		
Totals	**25,334**	**11,980**	**1,047**	**443**	**298**	**69**	**258**	**131**	**44**

employment had been on an almost ad hoc basis and, in consequence, was not as efficient as it could be. A major factor that helped to change this was the work of the Fighter Command Operational Research Section (ORS), one of the series of Command ORSs that played such a vital but often forgotten part in the development of the RAF's air power potential. In spring 1941 the Fighter Command ORS undertook a major survey into the state of AI development and employment which included a tour of the nightfighter squadrons and 'full and frank' discussions with the crews. The result was the production of a series of instruction manuals and the instigation of a training school for the Navigation (Radar) Leaders from the squadrons. The result of one such tour was reported in October 1941, the basic premise of the visits (to Nos. 29, 85, 219, 600 and 604

Squadrons) being to 'find out what operational use was being made of the apparatus'. The problems that the team perceived were summarised thus:

1. Operators do not use the best acknowledged methods, a number have had no proper instruction and those who have are not encouraged to continue with the best methods.
2. Many practice intercepts are perfect – but they are of little value because of 'cheating' by the pilot, generally unconsciously, and a lack of proper planning.
3. There is no officer in squadrons responsible for AI operations. In consequence, there is no proper organisation for the training and general employment of AI within the squadrons.

Each AI unit spent a great deal of time on practice intercepts, working with GCI stations to perfect all aspects of the system. However, as this report pointed out, there was a tendency, as most of these training sorties were flown in daylight, for the pilot to help out the radar operator with a spot of judicious aircraft positioning relative to the target. This and most of the other points raised in the ORS report were addressed, and the standards improved markedly during the early part of 1942.

Among the mass of paperwork being generated in an effort to refine the system, Fighter Command issued a 'Description of Operational Procedures' for GCI control of AI-equipped aircraft, which stated:

> After taking off a pilot will operate on Channel D and be controlled to his patrol line by the Sector Controller who will obtain occasional fixes of his position by means of voice fixing. Once on the patrol line the pilot should be able to maintain his position by an illuminated patrol line or by predetermined patrol courses and times.
>
> The Sector Controller will feed aircraft one by one from the patrol line to the GCI by giving the pilots vectors which will bring them to a convenient point for the GCI Controller to take over and operate them on Channel B. This feeding must be well practised so that the GCI Controller will not have to wait for another aircraft after putting one into AI contact with the enemy; nor should he have to bring an aircraft from a distant point before commencing the interception.
>
> It cannot be over-emphasised that only by good systematic co-operation between the GCI Controller and the Sector Controller can real use be made of a number of fighters in the air. The period during which hostile raids are crossing the area of the GCI is limited, and between them the Sector Controller and the GCI Controller must endeavour to secure the maximum number of AI contacts during that period.
>
> The GCI Controller is to put the fighter into AI contact with an incoming raid as quickly as possible. He will not attempt to control the fighter once good AI contact has been achieved. The AI operator is to 'flash his weapon'

> at extreme AI range and directly contact is made is to inform the pilot. Once good AI contact has been established the pilot will give 'Tallyho' and switch on to separate inter-communication. The rate of supply of fighters to the GCI Controller should be arranged so that one can be handed over, at a suitable feeding point, at least once every 10 minutes.
>
> Once in AI contact the pilot will endeavour to close and engage the enemy. When he has achieved this or lost contact he will return to Channel D and call the Sector Controller, who will then instruct him to return to his patrol line or to land.

This stated the perfect GCI procedure, one that in essence is still employed today, and it proved to be markedly successful; increasingly so as crews (and controllers) became more experienced and AI technology improved. Among the AI developments was the long-awaited pilot set for single-seat operations.

The Mk. VI, an auto-strobing set suitable for employment in single-seat aircraft, was being put into limited production late in the year, and was being fitted to Hurricanes and Typhoons for trials in December. Also in December the first of the nightfighter squadrons received the Mosquito, No. 157 Squadron collecting Mk IIs and having AI Mk. V installed. For some time concern had been expressed that enemy aircraft were getting away because of limitations in the performance of the Beaufighter (the same argument that had been used about the Blenheim). It had been decided early on in Mosquito development that the type would make an excellent nightfighter, but there had been many calls upon the Mosquito for other roles.

The RAF's tabular summary of night defensive operations makes interesting reading, and shows the changes made during the year (see Table 4). Note, for example, how the numbers of twin-engined and single-engined nightfighter sorties alters. Under RAF sorties, the first figure refers to multi-engined aircraft and the second to single-engined aircraft. The facet of most concern in these statistics was that the percentage of AI contacts resulting in combats had not improved. However, it was considered that recent changes, including the introduction of the Mosquito, would redress this. The prospects for 1942 looked reasonable.

It was during 1941 that the Admiralty at last took a serious interest in the problems of nightfighters, the primary cause being the presence of Focke-Wulf Condor aircraft shadowing convoys at night and relaying position information to aircraft and U-boats. Initial trials were made with Fairey Fulmars equipped with AI Mk. IV, followed by the formation of a nightfighter training unit and the search for a suitable operational nightfighter for carrier operations. However, it was to be some time before these early steps came to operational fruition.

In the closing weeks of the year, Goering made a decision that would put the Germans some way behind the Allies in the development of AI equipment. He rejected proposals to develop centimetric-wave sets, contending that the reflection factor would make them unusable. Similar interference in proposals for what were seen as defensive systems (bear in mind that in the early years Luftwaffe strategy and procurement was very much that of offensive systems) created huge difficulties for the Reich defence network.

The launching of the German assault on Soviet Russia, Operation Barbarossa, in June 1941 added yet another dimension to the air war. This bloody campaign was to prove very much a land battle; although strong air fleets were employed by both sides, it is very much a side-issue in this account. Much of the Russian air force was destroyed on the ground in the early weeks of the campaign, and those units that did engage the enemy found themselves greatly outclassed. Hence the reaction among the Russians was to conduct night bombing, although these attacks were small-scale and not very effective. August brought another threat to the German homeland. The first Russian night bombing of Berlin took place on the night of the 7/8th, although only one of the five aircraft actually bombed the target. Russian efforts were to remain small-scale for many months, but they had served notice that the night defences of the east could not be ignored.

The Luftwaffe bombers were somewhat more effective, and by July were carrying out a Blitz of Moscow. On the night of 21/22 July, four waves of bombers of KG3, KG53, KG54 and KG55, some 200 aircraft, attacked the city, dropping 104 tons of bombs and 46,000 incendiaries. The defences claim to have shot down 22 aircraft, 12 of these falling to cat's-eye nightfighters working with the searchlights. The technique was no different from that used by the RAF. General Gromadin had just under 800 heavy guns for the defence of Moscow, along with the newly-formed IVth PVO Fighter Corps of 600 aircraft, commanded by Colonel Klimov. Among the nightfighter assets were a number of Pe-2 and SB bombers that had been fitted with searchlights, in a similar concept to the RAF's Turbinlite. Among the fighter types employed on air defence were MiG-3s, LaGG-3s and -9s, Yak-1s, I-16s and I-153s, all of which at one time or another attempted nightfighter work. Some of the units had been formed specifically for night defence, including the 2nd Independent

Night Fighter Squadron, led by Andrei Yumashev and composed of leading test pilots from various organisations.

Throughout its history, the Soviet air force has encouraged the ramming of enemy aircraft as the supreme sacrifice in defence of the mother country. On the night of 7/8 August, Junior Lieutenant Talalikhin became a Hero of the Soviet Union when he rammed an He 111 near Moscow. He reported:

> I managed to hit the bomber's port engine and it turned away, losing height. It was at that moment that my ammunition ran out and it struck me that, although I could still overtake it, it would get away. There was only one thing for it – to ram. If I'm killed, I thought, that's only one, but there are four fascists in that bomber. I crept up under its belly to get at its tail with my propeller, but when I was about 10m away a burst of fire hit my plane and shot my right hand through. Straight away I opened the throttle and drove right into it.

The bomber fell to its destruction, but Tulalikhin was able to escape from his stricken aircraft. This was not a nightfighter tactic favoured by other air forces!

In like vein the Luftwaffe had to provide an element of nightfighter defence to supplement the flak guns. The first confirmed success on the Eastern Front went to Leutnant Rudolf Altendorff of 3/NJG3 on 20 October. Other units, such as the day fighters of JG54 around Leningrad, took part in moonlight operations with some success.

During 1941 the air war had increased in scale, complexity and ferocity. Night operations were established as an integral part of air strategy, and technology had provided the means of finding the enemy in the dark.

Chapter 5

THE NIGHT WAR HOTS UP

For both sides, 1942 was to be a year of success and failure, expansion and modification. It was a critical year for Bomber Command. Its supporters had yet to prove that the Strategic Bombing Offensive was having any significant effect on Germany, while its detractors continued to apply pressure for resources to be reallocated to other areas. In at least one respect there was no doubt that the Command's efforts had paid dividends. The Germans had been forced to devote much attention, manpower and resources to the problem of the defence of the Reich. By January 1942 the total number of flak guns had almost doubled, and the nightfighter system, with its associated radars and control networks, had been greatly expanded. However, in many eyes Bomber Command had yet to prove its value. February 1942 saw Air Marshal Arthur Harris appointed its Commander-in-Chief, and there began a period during which the Command was transformed into an effective and devastating force.

For the Luftwaffe defences, and the nightfighters in particular, it soon became obvious that the threat from the night bombers was taking on an increasing importance. This was to be the year in which certain German nightfighter pilots began to establish a name for themselves, even though some had come to the role very reluctantly. Hans-Joachim Jabs made his first night kill, a Stirling, in June. He had started his career on Bf 109s but quickly moved to Bf 110s with II/ZG76 (during which time he scored nineteen victories). He was transferred to nightfighters with II/NJG3 in November 1941, and became very frustrated by his lack of success, having scored only six night victories after a year. Meanwhile, other names were coming to the fore: Heinz Schnaufer, Prinz Heinrich zu Sayn-Wittgentstein, Helmut Lent, Ludwig Becker and Paul Gilder. All of these became nightfighter 'experten'; all became exponents of the multiple kill. Schnaufer was particularly successful in the latter regard. Having shot down his first bomber on 2 October 1942, he went on to achieve 121 victories.

In the first few months of 1942, Bomber Command concentrated on small-scale raids against German capital ships, the *Gneisenau* and *Scharnhorst* at Brest being particular 'favourites'. Meanwhile, new equipment, such as the Gee navigation aid, and aircraft, such as the Avro Lancaster, were transforming the Command's capabilities. All that remained was to return to the offensive against German industrial cities. During March, Essen, with its extensive Krupps munitions factories, was subjected to a number of medium-sized raids.

However, before the series of raids against Essen, the Command mounted its largest operation yet, with 235 aircraft attacking the Renault factory at Boulogne-Billancourt near Paris on 3/4 March. The major significance of this for the present study is the level of concentration (by time) of the bombers over the target: previous attacks had averaged 80 per hour, but this night the bombers achieved 120 per hour, the tactic being aimed at both increasing the weight of the attack and reducing the aircraft's exposure to the defences. Many 'experts' had feared that too great a concentration would bring about an unacceptable level of collisions, but this did not prove to be the case and concentrating bombers into a very short time span over a target became a standard Bomber Command tactic.

Essen was attacked on 8/9, 9/10, 10/11, 25/26 and 26/27 March (the 10/11 raid being significant as the first Lancaster attack on a target in Germany). Although these attacks were Gee-guided, they failed to achieve any level of concentration or accuracy on the target. The night defences continued to take a toll of the bombers, the final raid losing almost 10 per cent of the 115 aircraft taking part, the majority falling to night fighters along the inbound and outbound routes.

RAF intruder operations were now firmly established as a means of night defence, confirming the old military maxim of attack being the best form of defence. Not only did such operations reduce the weight of enemy bomber attacks on the UK, but they also provided important support for the RAF's own offensive bombing campaign by attacking the nightfighter system. The intruder organisation was still small-scale at the beginning of 1942, with No. 23 Squadron's Havocs and Bostons and the long-range Hurricanes of No. 3 Squadron. During January most activity comprised bombing and patrolling enemy airfields, Lille, Laon and Dinard-Pleurtuit being the main targets. February saw a number of operations, including two sorties flown on 26/27 February in support of Operation Biting, the commando raid on Bruneval. This had a distinct effect on the night war, as the aim of the commando raid was to learn the secrets of the German EW radars. The Parachute Regiment/Royal Engineer force, led by Major John Frost, not only succeeded in 'acquiring' parts of the Würzburg radar, but also brought back one of its operators. This provided vital information in the growing electronic war.

After the series of attacks on Ruhr targets, during which the nightfighters had become an increasing threat, many bomber crews reporting sightings and combats, 28/29 March saw Bomber Command launch its most effective raid of the war to date. The target was Lübeck, and 234 aircraft, mainly Wellingtons, were tasked in three waves, the first of which comprised experienced crews using the Gee navigation aid in an attempt to make the first bombs accurate, thus enabling subsequent waves to bomb the fires. Conditions were perfect, and extensive damage was caused while losses among the bombers were just under 5 per cent (twelve aircraft), the majority falling to nightfighters. However, for the remainder of spring the primary targets were once more in the Ruhr, although Hamburg was also targeted. In both areas the defences were strong and particularly active.

May 1942 brought the next major development in the night war. In an effort to demonstrate once and for all that the bomber was the decisive weapon, Harris was determined to gather together a bomber force of 1,000 aircraft for a single raid. He was in no doubt that:

> . . . the bombing strength of the RAF is increasing rapidly, and if the best use is made of it, the effect on German war production and effort will be very heavy over a period of 12 to 18 months, and such as to have a real effect on the war position.

The 1,000-bomber raid was to demonstrate the full power of a co-ordinated bombing plan. The target was the city of Cologne, and on the night of 30/31 May a force of 1,047 bombers, including 338 four-engined 'heavies', left their bases in Britain. Just under 900 are recorded as bombing the target. Losses were 41 aircraft, roughly half of which fell to nightfighters between the coast and the target (inbound and outbound), with flak claiming 16 and fighters four over the target area. Although this was the highest number of losses so far suffered in a single night, it was also by far the largest raid, and the loss rate was considered acceptable in view of the predicted level of damage inflicted upon the target. The raid had been supported by a number of intruder sorties against nightfighter bases, and although the Fighter Command Blenheims and Havocs had no luck (for the loss of two Blenheims), it was considered that this type of operation was vital to disrupt the nightfighter organisation. The other element that appeared to be in favour of the attackers was the sheer scale of the operation. The third wave had suffered the lowest loss rate, and it was

TABLE 5. THE 'BAEDEKER BLITZ', APRIL–AUGUST 1942

Date	Target	Sorties
April		
2/3	Weymouth	7
17/18	Southampton	44
23/24	Exeter	44 +
24/25	Exeter	?
25/26	Bath	163
26/27	Bath	83
27/28	Norwich	73
28/29	York	74
29/30	Norwich	73
May		
3/4	Exeter	90
4/5	Cowes	?
8/9	Norwich	76
19/20	Hull	132
24/25	Poole	96
29/30	Grimsby	66
31/1	Canterbury	77
June		
1/2	Ipswich	65
2/3	Canterbury	58
3/4	Poole	91
6/7	Canterbury	58
21/22	Southampton	94
24/25	Birmingham	80
26/27	Norwich	60
27/28	Weston-S-Mare	53
28/29	Weston-S-Mare	53
July		
6/7	Middlesborough	?
7/8	Middlesborough	53
25/26	Middlesborough	22
27/28	Birmingham	111
29/30	Birmingham	100
30/31	Birmingham	45 ?
31/1	Hull	?
August		
1/2	Norwich	?
4/5	Swansea	26
10/11	Colchester	?
13/14	Norwich	?
14/15	Ipswich	?
20/21	Portsmouth	?
26/27	Colchester	?

assessed that this was because of saturation and the exhaustion of the defences. According to German sources, only 25 nightfighters had been directed towards the bombers as only eight nightfighter zones had been penetrated. These few aircraft had done well – but much more could have been achieved with a more fluid system.

The 1,000-bomber raid was repeated against Essen (1/2 June) and Bremen (25/26 June), the latter also being significant for the first intruder operations by Boston and Mosquito aircraft. Although the three raids had not perhaps been the clear-cut demonstration of potential that Harris had intended, they certainly impressed many, including Churchill, and eased some of the pressure on the Command. This was just as well, as the '1,000-bomber' force had to be broken up to allow the large numbers of OTU aircraft to return to their vital training role. The demonstration had, however, also impressed the Nazi leaders, and greater efforts were demanded from the defence forces.

Following the Bomber Command attack on Lübeck, Hitler demanded 'Vergeltungsangriffe' (reprisal raids), a sentiment expressed by Goebbels in one of his speeches:

TABLE 6. DEFENSIVE RESPONSE TO THE BAEDEKER RAIDS, APRIL–AUGUST 1942

	Sorties	Destroyed	Probable	Damaged
April				
AI fighters	783	16	6	13
Cat's-eye	397	3	1	5
Turb/sat	3	1	–	–
Intruder	183	8	–	12
May				
AI fighters	547	14	3	5
Cat's-eye	187	–	1	1
Turb/sat	37	–	–	–
Intruder	198	10	2	7
June				
AI fighters	908	21	5	10
Cat's-eye	286	3	–	5
Turb/sat	68	–	–	–
Intruder	325	13	–	10
July				
AI fighters	1,231	36	8	20
Cat's-eye	655	36	8	20
Turb/sat	117	36	8	20
Intruder	309	9	1	9
August				
AI fighters	1,374	21	7	23
Cat's-eye	424			
Turb/sat	88			
Intruder	158	1	2	3

> Like the English we must attack centres of culture, especially those which have only little anti-aircraft defences. Such centres must be attacked two or three times in succession and levelled to the ground – then the English will no longer find pleasure in trying to frighten us with terror attacks.

The Luftwaffe bomber units were by no means a spent force, and in April they launched what was to become known as the 'Baedeker Blitz', the name arising from the chosen targets being British cities featured in the famous Baedeker Guides (see Tables 5 and 6). Bomber strength was built up to around 430 aircraft, about half of which were available for operations on any given night. The action opened on 23/24 April with a raid by 40 aircraft to Exeter, although in the event most actually attacked targets around Exmouth. The raid was scattered and not effective, with No. 604 Squadron claiming a Do 217. Exeter was the target again the following night, and Bath was targeted on the subsequent two nights, these latter raids being far more destructive.

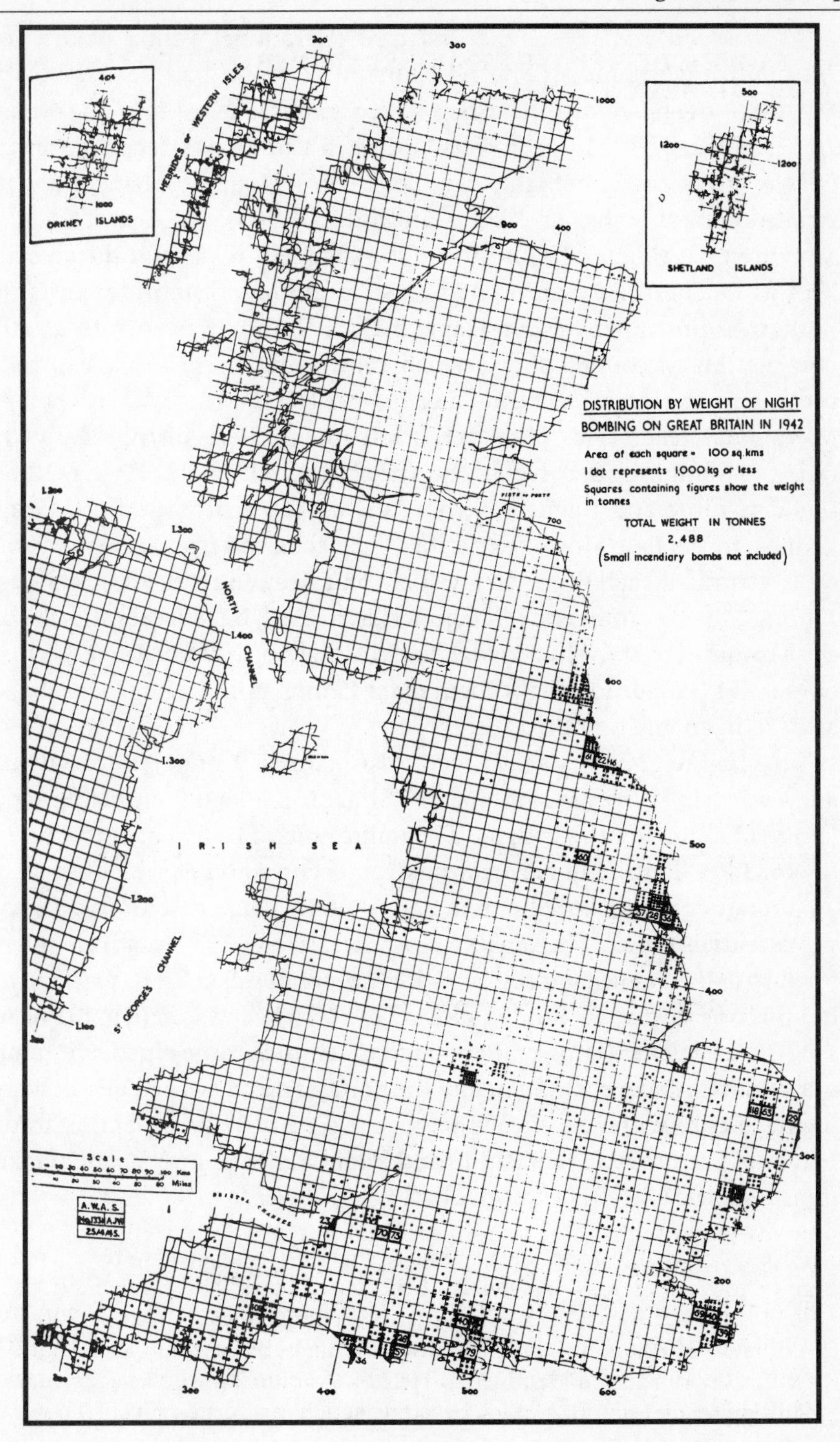
DISTRIBUTION BY WEIGHT OF NIGHT BOMBING ON GREAT BRITAIN IN 1942
Area of each square = 100 sq.kms
1 dot represents 1,000 kg or less
Squares containing figures show the weight in tonnes
TOTAL WEIGHT IN TONNES
2,488
(Small incendiary bombs not included)
ORKNEY ISLANDS
SHETLAND ISLANDS
HEBRIDES or WESTERN ISLES
NORTH CHANNEL
IRISH SEA
ST GEORGES CHANNEL
Scale
Kms
Miles
A.W.A.S.

This period was to see the first true operational outing of the new Mosquito nightfighters. The first unit to re-equip had been No. 157 Squadron the previous December, which acquired Mosquito IIs and had AI Mk. V fitted. By March they had fourteen aircraft on strength and were heavily involved with flight trials and sorting out problems, such as repositioning switches and lights and fitting blinds. April saw No. 151 Squadron's A Flight take on Mosquitoes, while B Flight, for the present, kept its Defiants. At their home base of Wittering, Cambridgeshire, the Station Commander, Group Captain Basil Embry, was keen to use the new aircraft to counter the German minelaying aircraft that had been proving so elusive. Thus began a series of PIs aimed at devising the best operational technique. However, such moves were disrupted by the renewed bombing offensive. In the early hours of 27 April, Flying Officer Graham-Little and Flight Sergeant Walters were airborne from Castle Camps on the first Mosquito patrol. They achieved two AI contacts but were frustrated each time when visual identification revealed 'friendlies'. During May the number of AI contacts increased, but success still eluded the Mosquito crews. That changed on the night of 29 May, when DD628 of No. 151 Squadron (Flight Lieutenant Pennington) claimed a probable against an enemy bomber.

The Fighter Night patrols were also extended and reinforced. An additional six Hurricanes of No. 253 Squadron joined the existing six at Church Fenton to work with the Turbinlite units. The first operational use of No. 1459 Turbinlite Havoc Squadron occurred on April 28/29, when an aircraft achieved an AI contact at 10,000ft range and at 4,000ft was visual with its target. The satellite Hurricane of No. 253 Squadron moved forward and, having good contact with the target, called for *no* searchlight, but this was turned on, illuminating a Do 217 which was held in the beam for 10-12sec despite evasive manoeuvring. However, the Hurricane could not get into a firing position before the bomber dived away. The following night, Flight Lieutenant D. Yapp was airborne at 0140 as parasite to the Havoc of Flight Lieutenant Winn and Pilot Officer Scott. The ORB takes up the account:

> Given vectors of 110 and told to climb to angels 14, at angels 8 were ordered to vector 060, on reaching angels 14 further vector of 030, handed over to Reclo. After various vectors of a northerly direction made contact on 340 at a distance of 10,000ft. Pilot Officer Scott brought them within 500ft of the bandit's port side and identified as He 111. As Hurricane had clear visual, Turbinlite not used. Bandit took evading action in a vertical dive, followed

> by F/L Yapp, firing from approximately 100 yards. Strikes observed as bandit dived and spiralled into cloud at 1,500ft at approx 400mph. Hurricane pulled out. Large fire seen below lighting up the cloud. He 111 claimed as destroyed.

Throughout July there were various successful contacts, most of which used AI to get into visual contact and so did not use the searchlight. However, the light was sometimes brought into use, as on the night of 12 August:

> AI contact at 5,000ft distance crossing port to starboard. On turning 40 to right, enemy aircraft was ahead and climbing. Hurricane given 'Warm' then 'Boiling' at 1,500ft. Havoc used searchlight flap and illuminated enemy aircraft but unable to identify it. E/A fired immediately on being illuminated, hitting Havoc in the starboard engine. Hurricane fired almost at same time, receiving return fire from the top turret. Thought to be He 111. Hurricane fired again, strikes seen on starboard wing, fuselage and engine.

This aircraft was claimed as damaged.

The figures in Table 6 bear out the general points that have been made throughout this study of RAF night operations; the limited value of cat's-eye fighters (good results if the conditions were favourable) and the consistent success of the AI aircraft. The balance sheet for the intruders reflected that of the German units, with losses being greater than confirmed 'kills'. However, such statements give no attention to the other credit side of intruder operations, namely the disruption of operations and the consequent reduction in the number of effective bomber or fighter missions.

> By the middle of 1942 the nightfighter force was exacting a steady and mounting toll of enemy raiders over this country . . . in view of the lack of replacements, either in aircraft or aircrews, for the west, they ensured that a heavy and sustained air offensive by the Luftwaffe against the UK was, temporarily at least, out of the question.

So said the RAF narrative in summary of the first part of 1942. While the German air effort against the UK was to continue until the end of the war, its scale and effect was now on a continuous downward slope. There were a great many factors in this; not only the loss rate to the RAF's night defences, but also the overall crisis within the Luftwaffe. Too many demands were being placed upon an organisation that was already overstretched; as the situation on the Eastern Front continued to deteriorate, so it was necessary to transfer units. Later, when the Allied day and night bombing offensive was devastating German industry and cities, priorities changed and greater effort was put into the nightfighter force. The Luftwaffe was 'firefighting' from 1942 onwards.

The three raids against Birmingham in late July had been in response to the Bomber Command raid on Hamburg (26/27 July), and were ordered by Goering in support of Hitler's reprisal raid policy. Only 60 bombers were involved on the first raid, no real concentration of bombing being achieved. The second night saw 100-plus bombers achieve better results, while on the third night numbers were down to 45. Fighter Command had mounted some 160 or so patrols each night, claiming seventeen destroyed (with a further six falling to the anti-aircraft guns). This equated to an overall loss rate of 10 per cent, a figure that the German bomber units could not afford: 'This was a satisfactory achievement by Fighter Command and may have been the deciding factor which governed the cessation of large-scale raids on this country.' It is certainly true that from this point onwards the numbers of raiders dropped dramatically. One of the largest raids in the latter part of the year was that on Canterbury (31 October/1 November), in which only fourteen aircraft took part. On this night Fighter Command flew 113 sorties and had an excellent night, No. 29 Squadron claiming four Do 217s.

Meanwhile the Luftwaffe was making changes at home and on the Eastern Front. The early months of 1942 saw a new German nightfighter undergoing operational trials, the Dornier Do 217J. This aircraft went into service with NJG1 and certain nightfighter units operating on the Eastern Front in Russia.

By mid-June the German southern offensive in Russia had reached the fortress of Sevastopol in the Crimea. The Luftwaffe pounded the city defences day and night, and there was little that the Russians could do in response. In an effort to supply the beleaguered city they flew a nightly supply run, and in the period 21 June to 1 July some 288 sorties were flown, running the gauntlet of nightfighters and flak.

Throughout the summer the Russians increased the scale of their night bomber operations against supply routes and troop concentrations. The Po-2 became the backbone of this operation and also proved to be an early 'stealth' aircraft, as its plywood construction proved hard to detect on radar. In an effort to counter these nightly raids the Germans created mobile flak and searchlight units, often railway mounted, and also employed mobile GCI sites to control the limited number of nightfighters.

In June a detachment of NJG5 under Oberleutnant Alois Lechner had moved to Dommgarten and Labian to work with rail-mounted GCI sites.

They achieved few successes. Earlier in the year a short-lived detachment of 2/NJG3 in the Smolensk area had highlighted a number of problems, and a subsequent high-level report suggested a major expansion of nightfighter resources – but in the meantime 'do the best you can'!

In what many have seen as one of the decisive turning points of the war, the German army was drawn into the bloodbath of Stalingrad. Having failed to take the city or prevent the Russians from feeding in reinforcements, the 6th Army was in a vulnerable position. The Russian counter-offensive was launched on 21 November, supported by massive firepower and over a thousand aircraft. Within days the German forces had been surrounded and forced to rely on air supply for their very existence. While they still held airfields within the Stalingrad pocket, the Luftwaffe flew day and night missions to keep the army alive and fighting. As the Russian net closed and aircraft losses mounted, so the night-time sorties became more important. Russian fighters flew round-the-clock air defence patrols. During the period 24 November to 31 January the Luftwaffe lost 488 aircraft, although a large number of these fell to Soviet anti-aircraft guns or were destroyed on the ground.

The Mediterranean and Western Desert theatres were also somewhat lacking in night defences. Both sides undertook night bombing operations, but these remained on a small scale, especially when compared with those over Europe, and thus attracted little in the way of response. Even though Bomber Command was undertaking bombing missions to the industrial cities of northern Italy, the Regia Aeronautica was slow in developing a response.

One of the primary nightfighters being used by the Italians was the Fiat CR.42CN (CN = Caccia Notturna – nightfighter), which had proved itself to be quite unsuited to the task. A request was placed for Germany to supply Bf 110s for this role, and agreement was reached for the delivery of 24 such aircraft to help counter the RAF night bombing of northern Italian industrial cities. However, initial deliveries stopped at only three aircraft, followed by an offer to supply the Do 217 instead. Although a number of Regia Aeronautica units had a nightfighter role (usually as an adjunct to their primary day fighter role), one of the few to specialise in this was 59 Gruppo's Stormo 41 (Nos. 232 and 233 Squadrons). In December 1941 they were based at Treviso, equipped with the CR.42CN and, at some stage in 1942, a solitary Beaufighter. It appears that the Beau had been captured on 1 January when it landed at Magnisi, Sicily. It was subsequently destroyed in a single-engined landing the following January. The unit was tasked with the night defence of Central and Northern Italy. Crews were sent to Germany for training on the Bf 110 and Do 217, as were crews from the other nightfighter unit, 60 Gruppo's Stormo 41 (Nos. 234 and 235 Squadrons). The first three Bf 110Cs to be transferred were renumbered MM964, MM1358 and MM1804, two of these going to No. 235 Squadron. Based initially at Treviso, this unit had a mix of CR.42CNs and Bf 110Cs, although towards the end of the year, when it was at Lonate Pozzolo, it also acquired Re.2001s. The major stumbling block for operations was the lack of AI radars; the Germans tended to supply out-of-date aircraft with no AIs.

A limited number of German units had moved to the area to bolster the defences in late 1941, including 1/NJG2 operating from Sicily. The first night kill was made on 13 December, when Oberfeldwebel Sommer destroyed a Beaufighter over Crete. During the early part of 1942 the unit's aircraft achieved a number of 'kills', but had little impact on the overall air situation, especially as the Allied air forces continued to build up both day and night capability. Meanwhile elements of NJG3 were operating out of Benghazi.

Middle East Command's ORBAT for 27 October 1942 showed only two nightfighter elements under its command; No. 89 Squadron's Beaufighters at Abu Sueir as part of 250 Wing (with a detachment at Luqa), and No. 46 Squadron's Beaufighters at Edcu as part of 252 Wing. However, in view of the situation, further changes were made, No. 23 Squadron moving to Luqa at the end of December. The first night intruder sortie over Sicily took place on 29/30 December by two aircraft, one of which was flown by Wing Commander Wykeham-Barnes. These nightly patrols continued into January 1943, and on 5/6 January Flight Sergeant Clunes in DD794 intercepted and damaged a Ju 52 over the western part of Sicily, the first successful intercept by the unit in its new theatre. However, in common with all the RAF intruder squadrons in this theatre, much of the effort was directed against ground targets, with lines of communication, and especially trains, becoming increasingly important. As one of the most experienced of all night operations units, No. 23 Squadron applied that experience to their new operational theatre.

The American entry into the war at the end of 1941 had extended the operational sphere across the Pacific region, an area in which air power, albeit primarily carrier-based naval aviation, was to play a major part. It was also a war in which the two major combatants, the USA and Japan, had given little thought to night operations. There is little to say for this region during 1942, although late in the year a number of night combats were reported over China.

A typical combat account is related in the Office of Naval Intelligence Weekly Summary for 25 November 1942:

> While flying a night patrol at 14,000ft at 0100 sighted three Japanese bombers passing over his home field, flying in V formation at 15,000ft. The fighter pilot approached from the dark quarter of the moon, but the bombers made a turn that put him up moon. Japanese gunners opened fire at 200ft, hitting the fighter and destroying its radio. He skidded quickly over behind the Japanese bomber and fired a burst at it from very close range, which set the Japanese bomber afire. He then took position behind the flight lead and from very short range also set this ship on fire. His engine then failed and he had to force land; meanwhile, the third bomber had been attacked and shot down by another nightfighter.

Although the RAF included night flying in its flying training, this was still fairly cursory and included little operational emphasis. The bomber OTUs in England had realised the need and were working hard to rectify the situation, but the fighter organisation was much slower to react. The build-up of 54 OTU as a night-flying training unit was to have a profound impact on the capabilities of crews. In May 1942, pilots (No. 21 course) and radio observers (No. 19 course) were using a mix of Blenheim marks for conversion to twin-engine operations and night flying, intermediate work and, on the Advanced Squadron, AI work. Pilots usually arrived from No. 12 Advanced Flying Unit (AFU) at Grantham and other crewmen from No. 62 OTU, Usworth, where they had conducted basic AI training on Ansons, using the AI Mk. IV.

No. 54 OTU was a major unit, and a typical month would see over 2,000hr flown, about one third being at night, with a very high accident rate, including all-too-frequent fatalities. By early the following year the OTU had 87 aircraft on strength (52 Beaufighters and 35 Blenheims). With the revised course it was hoped that by the time crews arrived on their squadrons they would understand all the basic principles of AI operation and the inherent problems of the nightfighter. This was a vast improvement on the previous 'learn as you go' routine.

On the operational front, AI aircraft were still in short supply, so the Turbinlite experiment continued. In late April, 11 Group instructed that: 'The state of Turbinlites in future will be two Turbinlite teams at readiness and the remainder of Turbinlite teams at 30 minutes' readiness.'

The trials and tribulations of No. 1452 Flight were typical of those that plagued each of the Flights. The number of operational patrols averaged only five or six a month, with Hurricanes being the most usual parasites, No. 1452 Flight having exchanged liaison with No. 264 Squadron's Defiants in favour of No. 32 Squadron's Hurricanes. Frustration among crews mounted as patrols continued to be notable only for their boredom; the ill humour was not improved by such incidents as the mid-air collision between Boston W8257 and its accompanying Hurricane on 2 June, with the deaths of the Boston crew. It has often been asked why the concept was kept in being after its apparent failure to show much promise. It must be borne in mind that part of the Fighter Command philosophy was to have sufficient force to counter any major German raid that was mounted against a single city. There was therefore a need to maintain a given size of nightfighter force (usually considered to be twenty squadrons), almost regardless of the type of aircraft employed. Had such a major German raid developed, it is quite likely that these 'hunter-killer' teams could have proved effective.

August brought a significant change, one that had been long requested by the Flights; the incorporation of their own fighters. Many Turbinlite proponents considered that the lack of such integration had been largely to blame for the lack of success to date. For No. 1452 Flight this meant the arrival of six Hurricane pilots. This was followed, in September, by the upgrading of the Flights to squadron status, No. 1452 Flight, for example, becoming No. 531 Squadron. On 10 September the squadron acquired seven Hurricane IICs from No. 32 Squadron, along with a few more pilots who, according to the squadron diary, were: 'rather a worry, for none of them were keen on the particular duty, and this squadron would probably have been happier with new pilots fresh from an OTU'. The expected improvement in the Turbinlite operation did not materialise, and although the units were kept reasonably busy it was eventually decided that they were no longer worth the effort; all were disbanded in early 1943.

The use of greater numbers of aircraft gave Bomber Command additional problems in getting aircraft safely to, and over, the target. It also prompted the tactical experts to look at new ideas. Two innovations had been employed during the 1,000-bomber raids – the use of a bomber stream and attempts to condense the overall time over the target, both means of saturating the defences in time and space. A tight stream of bombers, constrained over a particular geographic area and using height separation, would penetrate the fewest number of defensive boxes, thus reducing the chances of interception and exploiting the major limitation of the GCI boxes. The same basic argument applied to concentration over the target, as this would saturate directed flak (but not, of course, barrage flak).

As various German targets were attacked through the summer, bomber losses began to mount once more as the defenders amended their tactics to suit the changing situation. Much hard discussion culminated in the creation of the Pathfinder Force (PFF), an expert target-marking force seen as one solution to the problem of bombing accuracy. August also saw the first American bombing raid. From now on the defenders would be battered day and night, and fighter resources would be harder to come by. Despite all of this, Bomber Command was approaching crisis point. The ongoing loss rate had averaged 4–5 per cent, the non-sustainable figure, and morale had begun to suffer as crews calculated that their chances of survival were not promising.

While flak remained the greatest threat, crews did not fear it nearly as much as the nightfighter. A lucky direct hit from flak would destroy a bomber, but most of the time the flak caused varying degrees of damage, and it could be seen and heard. The ever-present threat of the stalking nightfighter, ready to launch a devastating directed attack, was a different matter. The following account by German nightfighter ace Wilhelm Johnen provides an atmospheric picture of such an encounter:

> 'Buzzard 10 from Berta. Enemy aircraft at 12,000ft. Course 280. Fly on course 100. Two couriers are entering your sector . . . Bank to port on course 280. Courier at your altitude. Give her full throttle.' At that moment I was ordered to slow down as I had already overshot the British machines. I throttled back and lowered my flaps to brake my speed. 'Enemy aircraft at 12,000ft on the same course, one mile to stern. Fly at 200mph and keep your eyes open.' Almost at stalling speed I let the adversary approach, keeping my eyes on the bright horizon to the north. At last I saw a small shadow ahead. I dived immediately and got below him. I was in no hurry and crept closer. The enemy bomber – a Vickers Wellington – was trundling wearily homewards. I aimed my sights on the enemy's port engine. The distance decreased – 150 . . . 100 . . . 50 yards. The rear gunner had already fired a few bursts, but he could not aim properly because his pilot was taking avoiding action. Now the bomber's wings were spread out against the northern sky as he went into a left-hand turn. At this moment I levelled my aircraft and let him fly into the cross-wires of my sights. I gave him a burst and the port engine was on fire. The fire appeared to have gone out so I made another attack. I grew impatient and rashly attacked direct from the rear. The rear gunner was waiting for me to approach and as we drew closer we opened fire at the same moment. As the burning bomber dived earthwards I noticed that my own plane had been hit. [Johnen, *ibid.*]

The year 1942 was to see reorganisation and expansion for the Luftwaffe's nightfighter force. May saw the creation of a new command structure intended to make XII Fliergerkorps less unwieldy. Despite these cosmetic changes, which did have some impact, there were still major problems: aircraft and equipment still had a low priority, partly because Kammhuber had managed to upset a large number of senior officers. Nevertheless, the overall effectiveness of the system had increased, and by August there had been a sharp increase in the delivery of Lichtenstein B/C sets, thus giving the crews more tactical flexibility.

Disruption and destruction of the German nightfighter effort thus grew in importance. The scale of RAF intruder operations rose to a peak in the summer, 336 sorties in June being the highest monthly total. These involved nine squadrons (Nos. 1, 3, 23, 87, 245, 247, 403, 418 and 485) who claimed 15 enemy aircraft destroyed and a further 10 damaged, and also made 169 ground attacks on airfields and other targets.

A typical month for the premier intruder unit, No. 23 Squadron, was July, when their Havocs, Bostons and Mosquito IIs flew the following missions (MY = marshalling yards):

1/2	4 aircraft. Evreux, St André.
5/6	5 aircraft. Evreux, St André, Caen, Amiens MY.
6/7	One aircraft. Avord. One Do 217 destroyed.
8/9	4 aircraft. Chartres, Beauvais, Evreux, Orléans. Squadron Leader K. Salusbury-Hughes shot down He 111 and Do 217.
11/12	2 aircraft. Evreux, Chartres.
12/13	6 aircraft. Caen, Rennes, Dinard, Evreux, Chartres.
14/15	2 aircraft. Rennes, shipping.
19/20	1 aircraft. Abbeville, Amiens.
21/22	5 aircraft. Chartres, Orléans, Rouen, Juvincourt, Creil.
23/24	6 aircraft. Chartres, Orléans, Tours, Juvincourt, Caen, Brétigny.
25/26	2 aircraft. Chartres, Orléans, Tours, Rouen MY.
26/27	1 aircraft. Juvincourt.
27/28	6 aircraft. Evreux, St André, Dreux, Chartres, Criel, Beauvais, Orléans.
28/29	8 aircraft. Chartres, Orléans, Amsterdam, Gilze-Rijen, Eindhoven, Soesterberg. One aircraft failed to return.
30/31	9 aircraft. Orléans, Brétigny, Château, Orléans, Avord, Serquex. One E/A shot down at Orléans.
31/1	6 aircraft. Juvincourt, St Trond.

Mosquito operations had been initiated on the night of 5/6 July, so the squadron found it particularly pleasing to score a victory on only the second night out.

DE HAVILLAND MOSQUITO

Without doubt, the de Havilland Mosquito was one of the greatest nightfighters ever employed, and its combination of adaptability and performance ensured its success in this role for many years. The prototype (W4050, the bomber variant) first flew in November 1940, and with the performance that the aircraft demonstrated from its earliest days it was obvious it would make a superb fighter. After various design changes, such as the addition of fixed armament, the fighter variant first flew in May 1941. As the Mosquito II this was given an AI set and became a trials aircraft. Operational service could not be long delayed, and the type went to the nightfighter squadrons in January 1942.

From this beginning sprang a whole range of Mosquito nightfighters that took advantage of improvements in the aircraft and, more especially, of the rapid development of AI sets. Centimetric AI VIII radar was fitted to the Mosquito XII, which entered service with No. 85 Squadron in February 1943; in May of the same year the first Mosquito XIIIs (with American radar) went to No. 256 Squadron. A further 99 Mk. IIs were given American radars and designated Mosquito XVIIs. 1944 saw the next nightfighter variant, the XIX, this type soon being employed on Bomber Support duties, hunting enemy nightfighters. This role was also performed by the Mosquito XXX, the last of the wartime series. However, the type was so good in the role that it continued to be developed postwar, with NF.36 and NF.38 variants entering service. The very last Mosquito to be built was an NF.38 (VX916) in November 1950. In due course the Mosquitoes gave way to Meteors, and a great many aircrew mourned their passing.

Mosquito Mk. XXX data
Powerplant: Two Rolls-Royce Merlin 76. Crew: Two. Length: 41ft 9in. Wingspan: 54ft 2in. Maximum speed: 407mph. Ceiling: 39,000ft. Armament: 4 x 20mm cannon.

The emphasis on German nightfighter airfields that became evident in the last few days of July was maintained over the coming months. By October virtually all effort was against such targets, with Leeuwarden, Deelen, Twenthe and Venlo being favourite haunts. The increasing number of squadrons operating in the intruder role, an average of ten squadrons for most of 1942, meant that a wider range of targets could be attacked. This was translated into attacks on road and rail communications, railway marshalling yards being high on the list. The most hectic month saw the intruders flying 169 sorties for the loss of 9 aircraft, along with the sighting of 63 enemy aircraft, resulting in claims for 15 destroyed and 10 damaged. Up to now the intruder operations had been somewhat ad hoc, but by late summer AOC Fighter Command had decided:

> . . . to centralise the control of intruder operations in a special operations room at Fighter Command Headquarters. By this centralisation I expect to achieve greater flexibility, better co-ordination in planning and timing of attacks, and an extension of the operational area.

The RAF history summarised intruder operations for the second half of 1942:

> The number of operations carried out has steadily diminished. This was owing to the small scale of enemy activity and to the weather. To give pilots experience, however, the long-range aircraft were sent deep into central France to operate against enemy night training centres. Compared with the successes earlier in the year, the results were disappointing. Between August and December, we claimed the destruction of only two enemy aircraft, although we claimed to have damaged 54. Our own losses amounted to 14 aircraft. The AOC-in-C has decided that during periods of enemy inactivity the intruder force should carry out offensive patrols over enemy territory by day as well as night. Should the enemy recommence large-scale attacks on the United Kingdom, the force would revert to its original role.

Throughout the period, the intruder role had been extended to other units at the request of Fighter Command; even Coastal Command was requested to task its Beaufighters with this role occasionally. The core of the intruder force remained the Mosquitoes and Havocs of No. 23 Squadron and the Bostons of Nos. 418 and 605 Squadrons, with Fighter Command's Hurricane squadrons playing a major role from time to time. Even the Typhoons of No. 609 Squadron were called upon to serve as part-time intruders during full-moon periods. Of the four main types involved during 1942, Havocs flew 206 sorties, Bostons 636 sorties (with the highest losses, 19 aircraft), Hurricanes 581 sorties and Mosquitoes 153 sorties. The effects of this campaign have already been alluded to, but, again, the RAF account states the case very well:

> The effects of intruder activity on the German Air Force are not precisely known but prisoner of war reports at the time showed that such activity caused the diversion of returning bombers either to waiting areas or alternate airfields, the adoption of special landing procedures and the use of illuminated decoys and dummy airfields. Enemy crews were therefore forced to operate with restricted facilities and at a higher than normal nervous tension. Again it is impossible to measure the effect of such tension but it seems likely that it impaired the efficiency of the enemy bomber force in some degree at least. It is probable, for example, that as a result of intruders, the enemy accident rate was increased.

At this stage of the war the intruder operations still had the dual purpose of reducing the German bomber force's offensive capability and countering the growing nightfighter force. As the months went by, the latter task would become predominant.

The increased use of radar and other items of equipment had turned the air war into an electronic battlefield, a situation that was recognised by 6

October, when the Senior Air Staff Officer (SASO), Bomber Command, chaired a meeting to discuss the adoption of RCM. The general conclusion reached was that the organisation of the German night defence system offered four possible targets for attack by radio countermeasures. These were:

(a) The early warning system, the radar components of which were mostly Freya, situated around the coastline of Germany and Occupied Europe, supplemented by further Freyas inland. All of these operated on frequencies of 120–130mc/s.

(b) The close control system at GCI stations, which was based on the use of two Würzburgs, one for plotting the fighter and one for the bomber. The frequency was around 570mc/s.

(c) The channel of communications between ground controllers and nightfighters. Instructions were passed by R/T in the 3–6mc/s band.

(d) Enemy AI operating on 490mc/s.

It was recommended that an effective airborne jammer be developed to counter the Würzburg, and that an airborne Mandrel jamming barrage should be established. In the interim, the temporary expedient of Shiver was introduced to provide some disturbance of the Würzburgs. The Mandrel jammer was primarily aimed at the early-warning radars and had been in service since late 1942. However, it was the more widespread use of the equipment, and the tactic of creating a Mandrel screen (with airtcraft flying a race-track pattern) that eventually produced results as an element within the overall RCM campaign. Various marks of Mandrel equipment entered service to counter different radar bands. Certain of the specialist squadrons were later given Shiver to supplement the Mandrel equipment, the Shiver system being intended to jam the GCI (Würzburg) sets and flak fire control radars. Steps were also taken to disrupt the fighter control frequencies, the earliest system being that of transmitting engine noise, using microphones placed in the engine nacelles, to blot out the controller's voice. This countermeasure, called Tinsel, was first used in December.

The Luftwaffe continued to send over night bombing raids, but these were usually small-scale (very few exceeded 40 sorties a night), though anti-shipping and minelaying activity remained high. The strongest raids were three made against the Birmingham area at the end of July. In a four-night period between 27/28th and 30th/31st of the month, approximately 205 German bombers attacked this area. AI nightfighters now had the major share of the defensive operations, flying 1,231 sorties in July, compared with 655 by cat's-eye aircraft. By September the proportion had risen to 1,086 AI against only 256 cat's-eye.

One of the greatest problems of all had always been that of distinguishing friend from foe. Although IFF Mk. I equipment had been introduced during 1940, quickly followed by the Mk. II set in an effort to overcome some of the technical troubles, it was by no means an ideal solution. Not until 1942 was agreement reached on the universal adoption of IFF Mk. III for British and American aircraft. It was intended that all ground and air radar sets be fitted with the equipment, which would interrogate a transponder on the 'target' aircraft. The task, requiring at least one million sets of airborne equipment, was immense, especially as it was hoped to have all aircraft fitted by the spring of 1943. To assist the GCI controller, a series of codes was used between the pilot and the controller: 'Make your cockerel crow' (switch on the IFF); 'Strangle cockerel' (switch off IFF); 'Canary please' (activate the G-band switch). The Germans were even further behind in this respect, and lack of an IFF system was to create severe operational restrictions for their nightfighters.

As far as development of the actual AI sets was concerned, 1942 was the year in which centimetric sets began to see more widespread use. As discussed in the previous chapter, sets such as the Mk. VII had been tested in late 1941 as a means of dealing with low-level raiders. Orders for this American SCR.520(UK) set had been placed pending further advances with the British systems for microwave AI. The Mk. VIIIA was an interim handbuilt set that enabled initial operational experience to be gained before delivery of the definitive Mk. VIII (1,500 sets produced by E. K. Cole). A number of variants of this set were produced and, in close co-operation with the TRE, most were in operation in the latter months of 1942. Nevertheless, the older sets shouldered the operational burden throughout the year. The overall improvement in interception statistics was in large measure due to increased proficiency among the crews, rather than to any simple equipment upgrade.

By late 1942 German bombing attacks on Britain had almost ceased:

> During these two months [October and November], only 199 enemy aircraft were plotted within 40 miles of the coast, excluding the Shetlands area. A further 59 aircraft were plotted in a more distant area outside the normal range of our night defences. The only serious raid during the period was against Canterbury on the night of 31 October/1 November. It was made by two waves of enemy bombers estimated at a total of 35 sorties, and of these, four were destroyed by AI nightfighters. Smaller raids were made on the East

> Midlands on the nights of 21/22 October and 24/25 October, consisting of eight and twelve enemy sorties respectively. In addition there were two small attacks on Tyneside on the nights of 11/12 October and 16/17 October by eight and seven aircraft respectively.

In response to the German activity, Fighter Command flew 124 sorties, all but four being by AI aircraft, resulting in eight combats and six enemy aircraft shot down (the other two combats ended in claims for a probable and a damaged). The AOC stated:

> I consider that 20 AI squadrons will suffice for all foreseen eventualities. This will mean that I will not require to replace the 2 squadrons originally sent out to North Africa. My proposal for the replacement of the Turbinlite squadrons is to add 6 intruder type aircraft to each of 10 AI squadrons. My object for recommending this reorganisation is to give more twin-engined aircraft for manning searchlight boxes in the event of a heavy attack on this country.

The latter part of 1942 also saw the Royal Navy brought on to the nightfighter scene, the Naval Staff having commented, in October, that they had reached: '. . . a stage where an operational development unit has become necessary. There is a need for an efficient naval nightfighter at a very early date. It is hoped that the Firefly will fulfil this need.' Naval interest in nightfighters had arisen in 1941, and trials had been conducted with the Fairey Fulmar equipped with AI Mk. IV. Promising initial trials had led to a consideration of carrier trials and the formation of a training unit. Although selected RN crews were attached to RAF squadrons to gain experience, it was not until June that No. 784 Squadron formed as the Nightfighter Training Squadron, equipped with six modified Fulmars. Strength built up only slowly, and the unit moved from Lee-on-Solent to Drem in mid-October. The following month brought further expansion with the creation of No. 746 Squadron as the Naval Night Fighter Interception Unit (NNFIU), the equivalent of the RAF's FIU. It was natural that the two should work together, so the NNFIU's Fulmars moved to Ford in December.

The Fighter Command summary for 1942 stated:

> For Fighter Command, the year was thus largely one of consolidation and unremitting patrol work. If there were no spectacular achievements, Fighter Command's success must be measured not by the number of aircraft shot down but by the relatively few occasions on which British towns and industries were troubled by night air attack.

This is a very valid point, both for the British and German defenders, and one that is often ignored when historians (or ill-informed parties) attempt to measure success by the number of aircraft shot down or own aircraft lost. Statistics do, nevertheless, play an important part in any summary. For Fighter Command the year saw 16,000 night missions and victory claims of 124 by AI fighters, 14 by cat's-eye, one by Turbinlite and 43 by intruders. German records show a loss of 205 aircraft to enemy action and a further 39 through other causes. As part of the continuous appraisal of the night defence system, it was decided to abandon the non-productive elements (such as the Turbinlite units, which took a great deal of the Command's night resources but produced little in return), in order to concentrate on the proven elements. For the RAF, the Mosquito/AI combination was to prove remarkably successful.

Although the Luftwaffe bomber forces had found 1942 a difficult year, their colleagues in the nightfighter force were establishing an effective organisation, largely due to the introduction of new equipment, but also through the adoption of a sound yet adaptable night defence system. Total strength stood at fifteen Gruppen, although not all were yet at operational readiness. One of the biggest problems – although this had yet to be fully realised – was the Allies' growing use of electronic warfare.

Chapter 6

TECHNOLOGY TAKES EFFECT

During 1943 the German night bombing of England continued to be low key, as it had been in the latter part of the previous year. As the year opened, Luftlotte 3 had a mere 67 serviceable long-range bombers, primarily the Do 217s of KG2 and the Ju 88s of KG6, although it was anticipated that this strength would increase over the next month or so. Nevertheless, these numbers would restrict any bombing campaign to an average of 30 or so sorties a night. Despite this, the German High Command was calling for a major assault by 'Angriffsfuehrer England' during the winter of 1943.

The low scale of German raids had led to a reassessment by Fighter Command of its night defences, and the deployment of assets to other tasks. This had come about as a combined result of the reduced threat and the crisis in the Bomber Command offensive that was being caused by the increasingly effective German nightfighter force. The RAF's nightfighters were taking on an increased intruder role in an attempt to redress the balance.

The RCM elements of night operations also saw major changes during 1943, one of the first new items of equipment being the Boozer warning device fitted to bomber aircraft. The system had been under trial with No. 7 Squadron's Stirlings in 1942, and an RAF air gunner described it thus:

> This gadget consisted of an external aerial fitted to the tail of the aircraft and a small panel of three lights above the pilot's instrument panel. There were two red lights and one white light. The red lights were for ground activity and the white light for enemy fighters. If a Luftwaffe 'sod' had your aircraft on his interceptor monitor then this light would flash a brilliant white. As soon as this appeared, evasive action of one sort or another without further ado. It worked for me twice – and I'm still here!

Reaction to the device among bomber aircrew was mixed. Many chose not to take immediate drastic action as the fighter might not have been looking at them, and to make violent manoeuvres in the bomber stream increased the risk of collision, but at the very least it gave a 'heads up' warning for the gunners to be even more vigilant.

January saw the abandonment of the Turbinlite experiment with the disbandment, on 25 January, of all the squadrons. The other major changes were to be the re-equipment of the intruder squadrons with Mosquito VIs, the establishment of intruder flights with Mosquito IIs on all AI Mosquito squadrons, and the continuing replacement of Beaufighters with Mosquito IIs and XIIs. The ORBAT for early January showed an overall strength of 30 squadrons, of which 10 were Turbinlite, with 390 aircraft, 60 being Turbinlite equipped. Mosquito strength had risen to 6 squadrons, each with an establishment of 16 aircraft.

At Drem, No. 1692 Radio Development Flight was formed to test a new homing device that was able to pick up transmissions from the Lichtenstein radar. This device, later to enter service as Serrate, was to add a new dimension to the night battle.

The only large-scale raid against England in January took place on the 17/18th of the month in response to a Bomber Command attack on Berlin the previous night. Some 118 German bombers approached in two waves, London being their target, but very few bombs fell in the target area. Fighter Command flew 119 defensive sorties and claimed the destruction of four Ju 88s and one Do 217, other aircraft falling to the anti-aircraft guns in what was another poor night for the Luftwaffe.

The bombers' usual tactics consisted of low-level, high-speed nuisance raids, against which the existing types of AI proved ineffective. However, the introduction of a limited number of Mk. VIIIA sets, initially with the FIU and No. 219 Squadron, provedotherwise. The first victory with the new kit occurred on 20/21 January, when an FIU Beaufighter shot down a Do 217. This success was followed up in the next few weeks to confirm the capabilities of the new radar: as a number of these interceptions were against evading targets at low level, it certainly appeared to hold promise.

The Luftwaffe planners considered that nuisance raids gave good results for little risk, so Fw 190s of SKG10 were employed from mid-April as 'high-speed' raiders. The speed of the attackers certainly provided a problem for Fighter Command, as interception had to be either very well planned or just lucky. The Mosquitoes of No. 85 Squadron were moved to West Malling to help counter the problem. A typical night's operation was that of 16/17 May, when seventeen raiders appeared over East Anglia and the South Coast between 2309 and 0425. The defenders claimed five – an impressive percentage.

Raids against Britain remained sporadic, the larger ones tending to be 'reprisals' for Bomber Command attacks on Berlin, primarily as propa-

Top: Douglas Bostons, represented here by AH525, joined the RAF's intruder operation.
Above: Wellington R1379 of No. 115 Squadron was shot down on 10 May 1941 by Leutnant von Bonin of II/ NJG1. The aircraft was forced-landed by the observer, Sergeant Bill Leg, and the crew were rescued by German flak gunners from a nearby site.
Right: A Dornier Do 17Z-7 of I/NJG2. (Manfred Griehl)

Right, upper: From 1942 onwards, single-engined fighter-bombers such as this Fw 190 were used for harassing raids by day and night.
Right, lower: A Beaufighter VI of No. 255 Squadron. The first aircraft of this type were delivered to the unit in March 1942 at High Ercall, and it took them to North Africa in November of that year.

Left: A No. 151 Squadron Defiant in early 1942, when the unit had a training role.
Below: No. 87 Squadron in its nightfighter role, Charmy Down, 1942. Note the black aircraft. The squadron also flew night intruder missions.

Left: A Defiant of No. 253 Squadron.

Right: Havoc BD112 of No. 23 Squadron. These aircraft were used for night intruder operations, attacking German airfields. The intruder campaign was originally aimed at bomber airfields to reduce the impact of the Luftwaffe's bombing campaign.

Left: Blenheim I K7159 of No. 54 OTU. One of the problems in the early part of the war was inadequate night flying training, and 54 OTU was formed to rectify this.

Right: In the absence of any American nightfighter types, a number of USAAF units were given Beaufighters. Here an aircraft of either the 417th or the 415th NFS is seen in Italy. (Peter Green)

Left: The nightfighter units in North Africa moved forward to keep pace with advancing ground forces. Here a Beaufighter of No. 255 Squadron spends its first morning at Souk el Arba, the most forward airfield in Tunisia, in December 1942. **Below:** On 29 March 1942 Avro Manchester R5830 of No. 83 Squadron is nicknamed 'The Pepperpot' after encountering a nightfighter on the Lübeck raid.

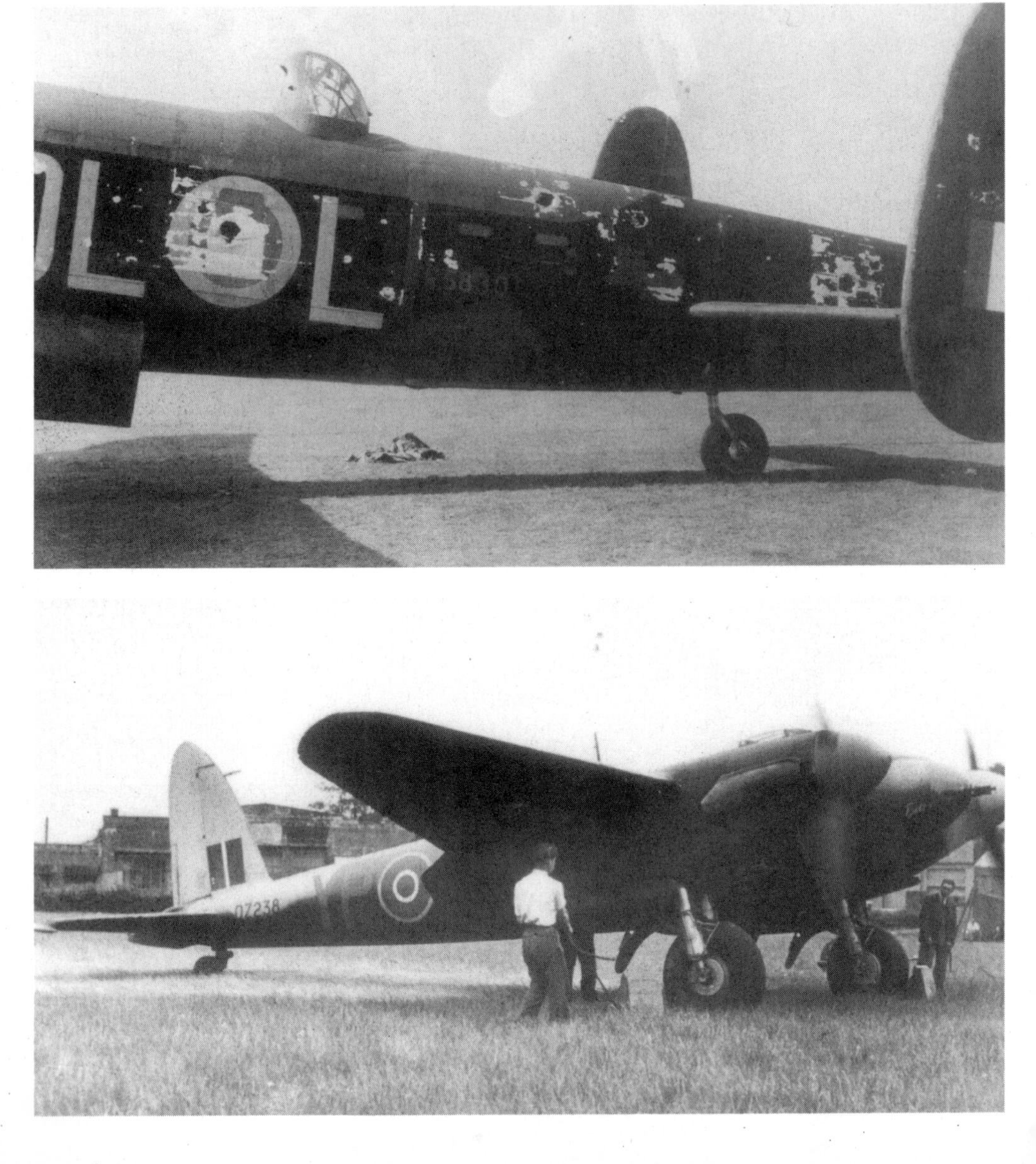

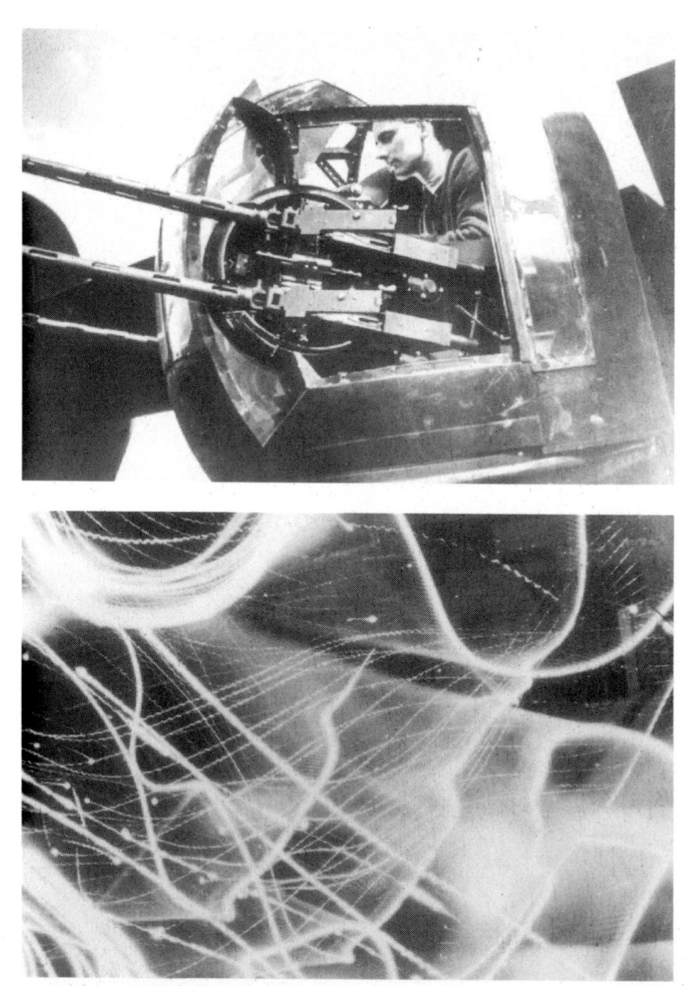

Opposite page, bottom: No. 23 Squadron was among the first units to receive Mosquitoes, acquiring Mk. IIs in July 1942. Seen here is DZ238.

Left, upper: A Halifax rear turret. The man isolated at the rear of the bomber had a vital task in keeping alert, spotting the nightfighter and preventing the deadly sneak attack.

Left, lower: A night photograph of the German city of Bremen, taken by an aircraft of No. 97 Squadron on 2/3 July 1942, shows the pyrotechnic display of flak and searchlights.

Below left: Stirling navigators plan their route, taking note of the latest intelligence to avoid the main flak zones. It was not so easy to plan a way around the nightfighters.

Below right: Squadrons of No. 325 Wing in North Africa record their victories on a piece of a German glider; from the point of view of this book, No. 255 Squadron's tally is of interest.

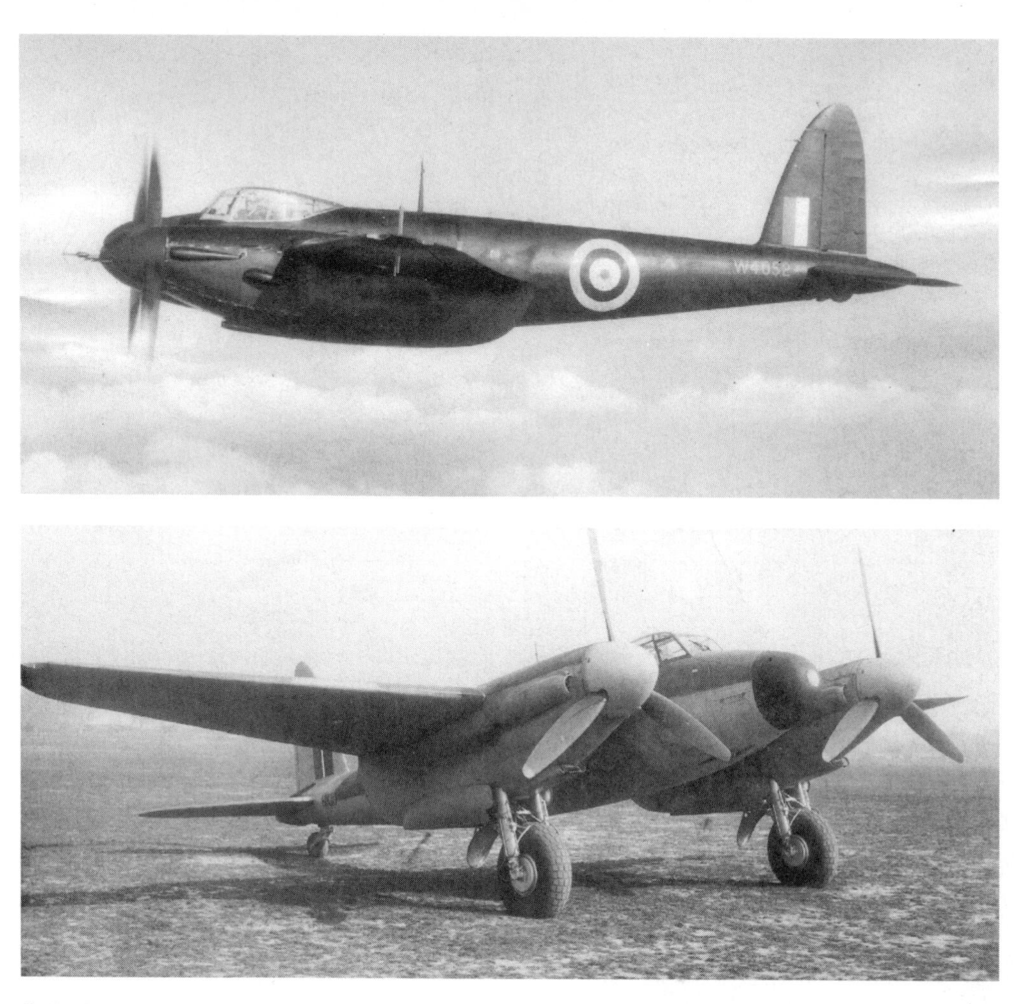

Left: Rearming an intruder Mosquito in Italy. The absence of night air-to-air targets meant that the units in this theatre were employed on night ground attack, primarily on lines-of-communications targets.
Top: Mosquito II prototype W4052. The need to improve nightfighter perfomance led to the early introduction of the Mosquito.
Above: The Mosquito NF.XII became one of the main nightfighter variants, proving particularly effective at countering low-level raiders.
Right: Oberst Helmut Lent, Luftwaffe nightfighter 'expert'.

Above: German nightfighter aces. Hauptmann Ludwig Becker (top left) was a nightfighter pioneer whose first victory was recorded on 16 October 1940 with II/ NJG1. By 26 February 1943 he had 46 night victories. He subsequently went missing on a daylight mission against American bombers. Major Walter Ehle (top right) started the war with 3/ZG1 but was kommandeur of II/NJG1 by October 1940. Before his death in a landing accident on 11 November 1943 he had achieved 36 night victories Oberleutnant Paul Gildner (above left) achieved four day victories on Bf 110s with 1/ ZG1 and was transferred to nightfighters with NJG1. On 24 February 1943 his aircraft suffered an engine fire that led to a fatal crash. His score stood at 44. Major Egmont Prinz zur Lippe-Weissenfeld (above right) was another high-scoring nightfighter pilot, achieved 51 victories before being killed in a flying accident in March 1944.

Top: A great *coup* for the RAF. Junkers Ju 88 D5+EV of IV/NJG3 defected to Britain from Aalborg on 9 May 1943, and was escorted in by Spitfires of No. 165 Squadron. It is now in the RAF Museum.
Above: The heavy bomber's blind spot. Although the Halifax carried the standard gun turrets, removal of the lower turret (not in fashion with the RAF) gave no protection from an attack from beneath.

Left, top: A Halifax IA displays the standard heavy bomber armament of a four-gun rear turret and four-gun mid-upper turret. Any attack from behind or above was well covered.
Left, centre: Halifax II L9485 carries rear, mid-upper, front and lower turrets, a total of twelve. 303 machine-guns providing all-round firepower, albeit of rifle calibre. Very few RAF bombers were equipped with lower turrets, armament being sacrificed to save weight and increase performance. This fact was exploited by German nightfighter crews.
Left, bottom: Lancasters of No. 83 Squadron line up ready for another night mission; how many will tangle with enemy nightfighters? By 1943 the night skies over Germany were very hazardous, and the chances of surviving a full tour of operations were becoming slim.
Right , top: The destruction of Essen. Given Bomber Command's capability for destruction, the German flak and fighter defences were of crucial importance.
Right, centre: Wing Commander J. R. D. Braham and Flight Lieutenant 'Sticks' Gregory, who together made up one of the great nightfighter crews.
Right, bottom: The British radar network, with its associated control and reporting system, played a crucial role both day and night. In the latter scenario the implementation of GCI allowed AI-equipped aircraft to achieve a high percentage of success.

Above: The Firefly NF.I prototype, DT933.
Right, upper: The FuG 220 aerials dominate the nose of this Ju 88G-1.
Right, lower: Crashed Ju 88C-6 4R+AS of 8/NJG2. The upward-firing pair of cannon ('schräge Musik') are clearly visible.

Above: Wilhelm Johnen was forced to land his Bf 110G-4, C9+EN, at Dubendorf after it was damaged in combat with a No. 35 Squadron Lancaster. **Right, upper:** A close-up of the tail of C9+EN, showing seventeen night victory tallies. **Right, lower:** The RAF Museum at Hendon has two German nightfighters on show, both of which landed in Britain during the war.

Top: A Northrop P-61A of Detachment B, 6th NFS, at Kagman Field, Saipan, on 21 June 1944. Having replaced the interim P-70, the Black Widows at last provided a workable nightfighter force.
Above: A P-61 of the 419th NFS at Henderson Field, Guadalcanal.
Below: The P-61 entered service in Europe with the 422nd NFS in February 1944, as part of the 9th Air Force, although it was June before operational status was reached. These aircraft from the 425th NFS are seen in France in 1944.
Right, upper: As it was the first purpose-designed American nightfighter, great things were expected of the P-61. In the event it entered service so late that air-to-air combats were few and far between.
Right, lower: Equipped with a pair of searchlights, the Firefly NF.II was another attempt at turning night into day using artificial light.

119509

Above: Vought F4U-2s of VF(N)-101 on board USS *Enterprise* in January 1944. (National Archives)
Left: RAF officers inspected captured Luftwaffe aircraft and equipment at the end of the war. (Via Chris Goss)

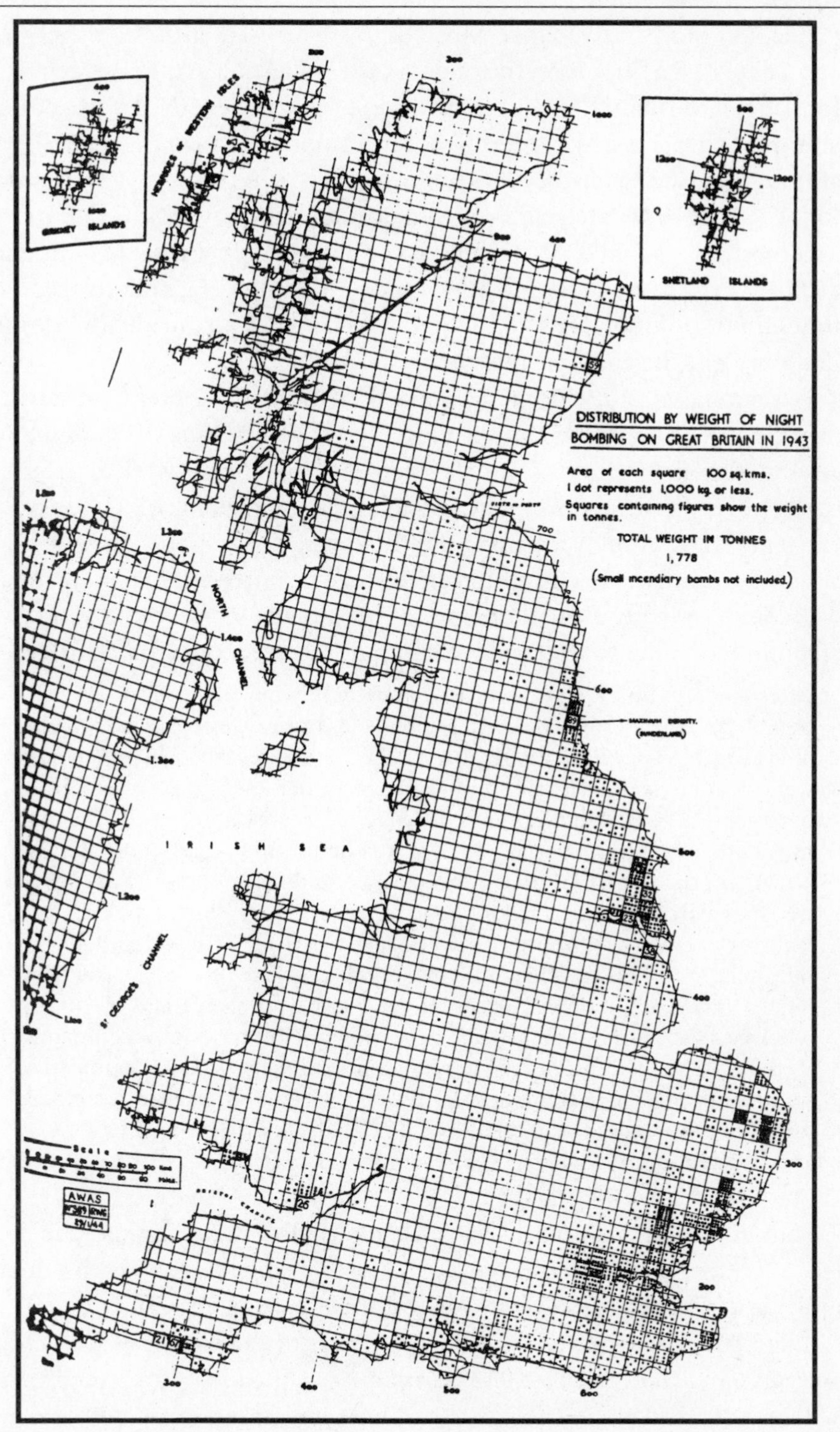
DISTRIBUTION BY WEIGHT OF NIGHT BOMBING ON GREAT BRITAIN IN 1943
Area of each square 100 sq. kms.
1 dot represents 1,000 kg. or less.
Squares containing figures show the weight in tonnes.
TOTAL WEIGHT IN TONNES
1,778
(Small incendiary bombs not included.)
ORKNEY ISLANDS
SHETLAND ISLANDS
HEBRIDES or WESTERN ISLES
NORTH CHANNEL
IRISH SEA
ST. GEORGE'S CHANNEL
MAXIMUM DENSITY.
(SUNDERLAND)
AWAS

ganda for the Nazi regime. Most of them caused little damage and, according to RAF sources, the raiders suffered an average 12 per cent loss rate. While the fighter-bombers mostly confined their activities to London and the South-East, the other bomber formations ranged farther afield, although coastal towns became popular targets. In the period up to mid-July 1943 the Luftwaffe had delivered 60 attacks against 10 cities, London and Norwich each having been attacked three times. Fighter Command claims for the same period totalled 65 destroyed and 11 probables, 34 of the definites going to AI Mk. VIII-equipped aircraft, an excellent score rate that bore out the early promise of this radar.

It was a similar story for the second half of the year. Fighter Command's nightfighter strength remained fairly static at 19 squadrons and 300 aircraft, though two units (Nos. 219 and 256 Squadrons) were transferred to the Mediterranean during the summer. On the night of 7/8 October the Germans first employed 'duppel', their version of the 'window' that Bomber Command had found so effective. The initial confusion was soon overcome, as many British operators had already worked in a 'window' environment. The RAF narrative summarises the second half of 1943:

> After the raids on Hull and Grimsby in early July the character of the night bombing changed from one of occasional raids by heavy bombers to sporadic attacks by high-speed bombers such as the Fw 190 and the Me 410; targets were found in southern and south-eastern coastal towns rather than in the north-east. The British night air defences were clearly more than a match for the enemy, for although the Do 217 was better equipped with defensive armament than other German bombers, it could not, because of its greater weight, easily evade nightfighters once they had marked it down. The two features of enemy night attacks in the latter part of 1943 were attacks by intruder aircraft on airfields in the Midlands and East Anglia in retaliation for the increasingly powerful combined bomber offensive and, secondly, renewed attacks on London by fighter bombers in the autumn, accompanied by the dropping of 'duppel'. From July to December 1943, the enemy flew some 818 long-range bomber, 165 fighter-bomber, and seven reconnaissance sorties overland – a total of 990. The RAF claimed destruction of 66.5 and the guns another 20.5. Once again the majority of interceptions were made under GCI control.

To put these figures into perspective, a typical Bomber Command raid of late 1943 was that to Berlin on 22/23 November, in which 764 bombers took part. The city was heavily hit and the attackers lost 26 aircraft.

The Luftwaffe nightfighter force underwent expansion, re-equipment and reorganisation during 1943 in response to the increasing weight and accuracy of attacks being delivered by Bomber Command. New units were ctions for use of this form in K.R. and A.C.I., ..., and War Manual, Pt. II., chapter XX., and ...A.F. Pocket Book.

No. of pages used for day ....

of (Unit or Formation) No 141 (Beaufighter) Squadron, R.A.F. WITTERING.

Place	Date	Time	Summary of Events	References to Appendices
	1943 June			
ering.	1st		More training at Drem, principally Air Firing. Two aircraftfrom Wittering on G.C.I Ptactice.	
	2nd		Trainiлg at Drem. Another re-equipped Beaufighter VI received from Defford.	
	3rd		Further training at Drem.	
	4th		Two aircraft from Wittering on G.C.I. Practice. Aircrews at Drem proceeding well with training.	
	5th		More aircraft ferried to and From Defford for fitting and on completion of re-equipping. Two aircraft from Wittering on day G.C.I. practice and two at night. W/Cdr Braham returned from Drem.	
	6th		All Squadron at Drem still training hard.	
	7th		W/Cdr Braham with Squadron Intelligence Officer and Special Signals Officer proceeded to fourth conference at Headquarters, Fighter Command.	
	8 & 9th		Training at Drem continuing well.	
	10th		Aircrew at Drem nearing completion of training. F/O MacAndrew on Fuel Consumption Test Flight, which proved satisfactory.	
	11th		Rest of aircrew returned from Drem to complete training at Wittering. A.O.C. No 12 Group, Air Vice Marshal J.O. Andrews, C.B., D.S.O., M.C. visited the Squadron and wished success in the new operations.	
	12th & 13th		Flying training proceeding at Wittering.	
	14th		First night of Squadron's Intruder Operations. W/Cdr Braham with F/Lt Gregory as Navigator Radio took off at 23.35 hours to patrol DEELEN, which they did uneventfully and at 02,10 hours whilst returning to the Dutch Coast saw an ME110 coming up behind. A dog fight in bright moonlight ensued in which W/Cdr Braham with the assistance of F/Lt Gregory completely out manoevoured the enemy aircraft which received a five second burst of cannon and machine gun fire. Strikes were observed on fuselage and port engine which caught fire. The ME110 dived vertically and crashed on the ground in flames N. of Stavoren. W/Cdr Braham then crossed the coast landing back at Wittering at 03.15 hours. At 23.35 hours S/Ldr Winn with F/O Scott as Navigator Radio took off to patrol EINDHOVEN. The patrol proved uneventful and the aircraft returned to base, landing at 02.40 hours. F/Lt. LeBoutte with P/O Parrott as Navigator Radio took off at 23.45 hours to intrude over GILZE. Apart from a moderate amount of flak when crossing the Dutch Coast, nothing of interest was seen and the aircraft returned to base at 03.15 hours. F/O MacAndrew took off at 23.45 hours to patrol in Holland returning at 03.25 hours, having seen nothing of interest. F/O Brachi with P/O MacLeod as Navigator Radio took off at 00.03 hours, but shortly after crossing the Dutch Coast was forced to return with instrument trouble and landed at DOWNHAM MARKET at 01.30 hours.	Form 541 Page 1. Form 541 Page 1. For m 541 Page 1.
	15th		More intensive training, but weather not suitable for operations. F/O K.J. Pamment with Sgt. Elliott and F/O H.E. White with his Navigator Radio P/O M.S. Allen rejoined the Squadron after being away for four months on ferrying duties to the Middle East. F/O White and P/O Allen proceeded to Drem on course.	
	16th		A suitable night for Intruder operations and six aircraft took part. S/Ldr. Winn with F/O Scott Navigator Radio, took off from Coltishall at 23.50 hours to patrol GILZE, making landfall at 00.23 hours. Nothing of interest seen except for flak, one near burst of which put all instruments out of action and aircraft returned to base at 02.45 hours. F/O MacAndrew with P/O Wilk Navigator Radio took off at 23.55 hours to intrude over EINDHOVEN. Patrolled target uneventfully until 01.15 hours when he saw an ME110 flying South and gave chase. The enemy aircraft appeared to be unaware of the Beaufighter's presence. At 01.20 hours F/O MacAndrew dived on him and getting into position gave a	Form 541 Page 1. Form 541 Page 1.

Extract from No 141 Squadron's Operations Record Book, June 1943

created, but wastage of crews and aircraft remained high, partly through the insistence of some commanders that the nightfighters should also take part in repelling the American daylight raids. They were no match for the agile American fighters, and irreplaceable experienced crews were lost for little good reason. One such loss was that of Hauptmann Ludwig Becker of NJG1 on 26 February: he had achieved 44 night victories.

The German position was not as simple as many postwar studies suggest. There was a great deal of in-fighting within the various elements of the Nazi military machine. Goering held a significant planning conference on 22 February. Among the many comments that came from this conference, the following are relevant to this study.

Discussing Kammhuber's insistence on production of the Fw 154 nightfighter, Goering said:

> If it were left to him, only nightfighters would be built and nothing else. He is emphatically egotistical about it. He says to himself, 'If no one else can use this aircraft then my allocation will be secure.' That is a point of view that should not carry any weight with us. If the aircraft is not superior to the Ju 188 then it will not be built.

Jeschonnek said:

> I would like to add something about this nightfighter [the Ju 188]. It has been laid down that sufficient aircraft are produced to provide us with 18 complete nightfighter Gruppen; but at the moment we can only put into operational Gruppen between 24 and 28 aircraft. Our ultimate objective is to have 24 Gruppen at full strength by 1 July 1944. There won't be sufficient quantities in 1943 to attain our targets.

Milch stated: 'The Ju 188 will definitely be one of our best nightfighters for a long time to come'.

On 18 March at Karinhall, Goering once more addressed his commanders:

> I now have to produce the means whereby at least some kind of counter-stroke may be delivered in view of the constantly increasing number of British bombers. Do not deceive yourselves, gentlemen; the British will carry out attacks with an ever-increasing number of these slow four-engined 'crates' which some of you hold in such contempt. He will deal with each and every city; he can navigate to Munich or Berlin with the same precision. Nothing bothers him. The nightfighters are successful on some occasions and unsuccessful on others. The flak can only play a defensive role or have a deterrent effect. The equipment with which they have to navigate and hit the target even in bad weather is ideal, while our instruments are always going wrong so that the nightfighters are always coming to grief and cannot do much about it.

No doubt many in Bomber Command would have been heartened to have heard these comments.

The 'untouchable' Mosquitoes continued to cause the German leadership great ire, and much effort was expended in trying to devise a solution. Operating out of Mainz-Friheim, III/NJG5 specialised in Mosquito hunting, although success eluded them until the early hours of 21 April, when Hauptmann Paul Zomer, flying Bf 110G-4 D5+BS, shot down one of the intruders. Another 'Mosquito-hunter' unit was II/NJG1 with a number of Fw 190s based at Hangelar in the Cologne area.

The *Bomber Command Quarterly Review* often included references to nightfighter tactics and engagements that had taken place. The issue for April–June 1943 included a feature on 'Encounters with Night Fighters', which stated in part:

> . . . the enemy High Command has gradually built up a truly formidable defensive front against Allied air attacks, and it is only because our own tactics and equipment have developed at least as fast that we can continue to operate successfully and on a larger scale than before. In spite of the

deepening of the controlled-fighter zone, which now extends in a solid belt from the Ruhr to the Dutch and Belgian coasts, our losses during the past quarter show no appreciable increase as compared with last year. Reports from air crews indicate that nightfighters frequently give up the attack without firing a shot if they see that the bomber has discovered them. It is very seldom that a crew that carries out the correct defensive procedures comes off worst. During the April–June period some 50 nightfighters were driven off in a seriously damaged condition, while 55 are considered to have been completely destroyed.

There were two distinct schools of thought among the air gunners (and the rest of their crew). Having seen a fighter, should you fire at it, thus betraying your position, or was it best to fire at everything to show that you were alert and thus not easy prey? Both methods were employed, but the latter was the official attitude and most seem to have supported this approach.

The *Review* went on to describe a number of recent engagements:

Wellington 'T' of No. 431 Squadron, returning from the raid on Stuttgart on the night of 14/15 April, observed a square of four white lights on the ground some 60 miles north-west of the target. These lights changed to a single line of white lights when the Wellington had passed, as if indicating its track. Presently the wireless operator, who was in the astrodome, reported an Me 110 at about 400 yards' range on the starboard quarter. As the bomber turned in towards the attack the rear-gunner fired a short burst. The enemy broke away immediately without firing and, passing underneath, came in again on the port quarter, opening fire at 300 yards. Meanwhile our aircraft was corkscrewing, but the rear gunner was able to reply with an accurate short burst, hitting the Messerschmitt in the nose. This was followed by a longer burst which caused the fighter to explode with a brilliant flash. Meanwhile a second Me 110 attacked from astern and underneath, fired a short burst which missed, broke away to starboard. As the intercom was u/s, the rear gunner signalled these attacks to the pilot by means of the call light. The enemy came in again on the starboard quarter but broke off without firing at 200 yards, when the rear gunner gave him a short burst. The Wellington was now flying at a very low level as the pilot had dived repeatedly to avoid attacks from below. But the Messerschmitt attacked once more, on the port quarter, firing a long burst that passed over our aircraft. At 200 yards' range the rear gunner fired and observed some hits on the Hun's starboard engine. As he passed overhead flames were seen coming from the nacelle of this engine. There was slight damage to the Wellington's port wing.

Many air gunners went through a tour without seeing an enemy fighter, while others had to fight with three or four in a single night. Sometimes they were presented with 'gift' shots. Tom Treadwell was a bomb aimer on Halifaxes with No. 77 Squadron, and on one particular night they had just released their bombs over the target when they noticed another Halifax off to port. The crew kept a close eye on the other aircraft, as they were all aware of the danger of being hit by 'friendly' bombs and wanted to keep well out of the way. At that point almost everyone on board saw an Fw 190 cruise past the nose of their aircraft and start to move into position on the rear quarter of the other Halifax. Tom recalls that it all seemed somewhat surreal and that no-one called the fighter until everyone suddenly shouted at the same time and the mid-upper gunner, who now had a perfect short-range side-on shot at the Fw 190, decided to let loose. Throwing all of the rules out of the window, he fired almost every round (3,000 plus) at the nightfighter. It went down almost vertically, but as there was a high cloud cover no one saw it hit the ground and it could not be claimed as a kill.

The shortage of aircrew and equipment for the Luftwaffe nightfighter force led some to reconsider the employment of single-engined fighters for night fighting. One of the strongest supporters was Major Hans-Joachim ('Hajo') Hermann, and in June it was agreed that JG300 should operate as 'cat's-eye' fighters in conjunction with the searchlights. In September this was expanded into a three Gruppen Division under Hermann – JG300 (ObserstleutnantKurt Kettner) based at Boon-Hangelar, JG301 (Major Helmut Weinrich) at Munich/Neubiberg, and JG302 (Major Manfred Mössinger) at Döbertiz. The Bf 109Gs and Fw 190As were fitted with exhaust flame dampers and anti-glare screens, but other than that were standard day fighters. All of this was a direct result of a conference chaired by Field Marshal Milch on 6 July 1943, during which Major Hermann presented his views:

> The subject in question is the employment of single-engined aircraft on night fighter operations. I do not think we are doing anything new, but are merely rather belatedly imitating the British. I consider the prospects of successful operations by this type of aircraft to be particularly good as it is highly manoeuvrable. There is also the fact that night fighter production is not making good progress and an effort can therefore be made to cover this time-lag in subsequent years by using instead single-engined aircraft which are available in fairly large numbers. I made this suggestion to night fighter control six months ago, but it was rejected. However, I did not give up my plan to use single-engined aircraft at night and watched the attack on Berlin on 1 March, following the action closely at the Flakdivision's battle headquarters. I discussed the matter with in detail with the Flakdivision on 8 April, and suggested to General Weise that I be allowed to begin an

experimental flight operating over Berlin. Permission was granted. I was promised four aircraft which I would lead on operations from a base at Berlin. The first operation took place on 20 April. This was an attack on Stettin, during which Mosquitoes flew on to Berlin. Unfortunately, the Freya installations did not locate the Mosquitoes at the exact time so that the order to take-off was delayed. The Mosquitoes were over Berlin at 6,000m by the time my aircraft had reached 5,000m and contact was of course not made. After only twelve days it was evident that this was not the right way.

For some time there was no contact with the enemy over Berlin and I therefore took my aircraft to western Germany. Three days ago the first contact on any scale was made during the major attack on Cologne and, in particular, on Muelheim. It was my estimation that about 500–600 enemy aircraft, including twin-engined types, were involved. This number of aircraft is very high. We had twelve aircraft in the air and the only arrangement we made was that we would co-operate with local flak. Cologne was not included in this arrangement. However, an attack was then made on Cologne. I therefore hurried to Cologne and waited over the target. After the first contact with the enemy I was surprised that the flak did not cease firing at enemy aircraft picked up by searchlights. Our pilots pressed home their attacks splendidly! They even flew through flak to approach the enemy aircraft. This was a completely improvised operation.

JUNKERS Ju 88

Designed primarily for the fast bomber/dive-bomber role, the Ju 88 proved to be an excellent weapons platform with a reasonable all-round performance. It was perhaps natural, therefore, that it was adopted as a nightfighter. The prototype Ju 88 was first flown in December 1936 and, despite early problems, soon showed promise as a bomber/dive-bomber. With the urgent need for nightfighters, the Ju 88C was employed both for night defence and night intruder operations, although it was 1942 before the first radar sets were fitted.

These aircraft were soon giving excellent service in the nightfighter units, but, as faster bombers were used by the RAF and Mosquito nightfighter escorts were provided, losses began to mount and successes diminished. The improved Ju 88G series were thus brought into the nightfighter role. Throughout the life of the Ju 88, a diverse selection of electronic equipment and weaponry was employed; at times it was hard to identify a 'standard' Ju 88 nightfighter. As part of the continued development programme, the Ju 388 was put into production as a specialised high-altitude nightfighter, but very few had been built by the end of the war. It would have been an awesome aircraft, with an armament of two 20mm and two 30mm cannon plus a variety of machine-guns.

Ju 88G-1 data

Powerplant: Two BMW 801D-2. Crew: Three. Length: 47ft 8in. Wingspan: 65ft 7in. Maximum speed: 356mph. Ceiling: 29,000ft. Armament: Various combinations of cannon and machine-guns.

> The aircraft could be clearly identified first as silhouettes, then by the exhaust; with the small single-seater aircraft it was possible to bank quickly and keep on the enemy's tail without difficulty. We flew five Fw 190s and seven Me 109-G4 and G6. The pilots with the 109s fly this type at night in preference to the 190; it is rather more manoeuvrable, especially at high altitude. At first we flew in echelon formation up to an altitude of 10,000m so that we could come down on the enemy very quickly. Later we went down from this altitude to between 4,000 and 6,000m and flew circuits around the target so that we might see the enemy aircraft more easily. The remarkable thing about this [type of] operation is that there should not be a limited number of aircraft in the target area; instead, any number of aircraft should be put in – the more the better – so that all the more enemy aircraft may be shot down. We made contact and fired on enemy aircraft sixteen times and made contact without firing seventeen times. For this first operation I had given specific instructions that pilots were not to open fire unless at first a correct target approach had been made.

The 'Wilde Sau' (Wild Boar) technique had proved its value, and was to be a major element of the nightfighter force for the next year or so. The bombers' ability to create major fires at their targets now unfortunately provided patrolling nightfighters with a useful source of illumination. If they lurked at the edges of the 'illuminated area' they could quite clearly pick up their bomber victims.

When Ju 88R-1 D5+EV of IV/NJG3 landed at Aberdeen/Dyce on 9 May, the RAF acquired the latest Luftwaffe nightfighter technology. The defection of Herbert Schmidt and Paul Rosenberger provided the intelligence services with a wealth of information on which to base new electronic countermeasures. The aircraft itself was soon being used for comparative performance trials (it now forms part of the aircraft display in the Battle of Britain Museum at Hendon). One event such as this could have a major impact on the course of an air campaign, and an analysis of the aircraft confirmed much that was already assumed but also added vital new details, especially regarding radar performance and capability.

Summer 1943 saw the Heinkel He 219 on operational trials with I/NJG1. This aircraft owed its origins to a Heinkel design of mid-1940 which had been seen the following year by Kammhuber, who considered that it had the potential to be an effective nightfighter and so pressed for its production. The type made its first flight in November 1942, but then underwent a number of changes which slowed down its development. Nevertheless, progress was made and in due course a number of aircraft were issued for operational trials. One of NJG1's 'experten', Werner Streib, proved the effectiveness of the type (flying He 219A-0 G9+FB) by

HEINKEL He 219 UHU (OWL)

Many commentators have described the He 219 as the best nightfighter of the Second World War. Although the design dated from early 1940, it was not until the following year that any attempt was made at further progress, the delay being largely due to political in-fighting among the aircraft manufacturers; Heinkel was not a fighter builder, and so should not build fighters. The first flight was made on 15 November 1942, but progress was still slow. Deliveries at last began in May 1943, the famous NJG1 being the first recipients, but of only a small number of aircraft. Despite early success there still appeared to be an element of official disinterest, and the Uhu only slowly came into full service. Production remained far slower than predicted. In common with most German combat aircraft, the type appeared in a wide range of sub-variants. By January 1944 only 26 aircraft had been delivered, and by the time production increased the nightfighter force was in decline.

He 219 data

Powerplant: Two Daimler-Benz DB603E. Crew: Two. Length: 53ft 7in. Wingspan: 60ft 8in. Maximum speed: 363mph. Ceiling: 32,150ft. Armament: 4 x 20mm plus 2 x 30mm 'schräge Musik'.

destroying five RAF bombers on the Düsseldorf raid of 11/12 June. However, for a variety of reasons (including political machinations and outright favouritism) the new aircraft was given no priority and only slowly entered squadron service. Nicknamed 'Uhu' (Owl) because of its owl-like front end, the He 219 has been proclaimed the best nightfighter produced in the Second World War, though this is a little hard to support because the type saw such limited operational employment.

Other aircraft were also under development. July 1943 saw the first flight of the Focke-Wulf Ta 154, designed to a 1942 specification for a two-seat nightfighter armed with four guns. Kurt Tank's design was all wood, and following the successful first flight an order for 280 aircraft was placed. The production aircraft flew in June 1944 and looked promising, but disaster struck when the only factory manufacturing the special glue used in the Ta 154's construction was destroyed in a bombing raid. This, plus a decision to concentrate production on existing successful designs, led to the cancellation of the project.

Meanwhile, on the night of 14/15 June the RAF introduced a new element into the night war; nightfighters equipped with the Serrate equipment designed to home on to the transmissions of the Lichtenstein radar. Five sorties were flown on that first night, and one Bf 110 was claimed as destroyed, the offensive aircraft having patrolled the airfields at Deelen, Eindhoven and Gilze-Rijen. Number 141 Squadron's ORB records the night's events:

> First night of the Squadron's intruder operations. Wg Cdr Braham with Flt Lt Gregory as navigator radio took off at 2335 to patrol Deelen, which they did uneventfully and at 0210 hours whilst returning to the Dutch coast saw an Me 110 coming up behind. A dogfight in bright moonlight ensued in which Wg Cdr Braham with the assistance of Flt Lt Gregory completely outmanoeuvred the enemy aircraft, which received a five-second burst of cannon and machine-gun fire. Strikes were observed on fuselage and port engine which caught fire. The Me 110 dived vertically and crashed on the ground in flames N of Stavoren.

'Bob' Braham was the most successful intruder pilot of all, an outstanding nightfighter pilot who, with his usual navigator/radar operator, Flt Lt 'Sticks' Gregory, proved the potential of Serrate. He had served with No. 29 Squadron, rising to be a Flight Commander, initially moving on at the age of 22 to command No. 141 Squadron. The Squadron Intelligence Officer, Buster Reynolds, said: 'Bob Braham's 100 per cent dedication and commitment to the task, whatever it might be, set him apart from other people and lesser mortals'. Even when he was posted to a ground job, on the night operations staff of 2 Group, he managed the odd trip – and added to his score – until he was shot down over Denmark in June 1944. His score stood at 29 confirmed, and his endeavours had been recognised by the unique award of a DSO and two bars, plus a DFC and two bars.

Fighter Command ORS report No. 494 examined the results of Serrate operations for the period 14/15 June to 6/7 September. During that time there had been 233 sorties giving rise to 1,180 'contacts', with 108 converted to AI pick-ups, of which 33 went to visual and 20 to combat, resulting in claims for 13 destroyed, 1 probable and 4 damaged. The report concluded:

> . . . the Serrate equipment is so sensitive in recording changes in azimuth and elevation of the target that the crews prefer to use it if possible for the whole of the interception, using the AI as little as possible for occasionally checking the range.

However, it also highlighted the fact that the poor performance of the Beaufighter was in large part responsible for many of the contacts not being converted into combats. Although the Mosquito was operational within the UK air defence squadrons, there were still not enough of these exceptional aircraft to go round.

The need to reduce the effectiveness of the German nightfighters had led to a number of suggestions for attacks on their home bases. While Serrate was, perhaps, the ultimate, although it would soon be employed more as an escort/sweep technique for the bomber stream, it was considered that Fighter Command aircraft should be able to undertake intruder operations. Under the codename Flower, the Mosquitoes of No. 605 Squadron flew the first such operations on 11/12 June. This was to prove very much a 'growth industry', and before long major efforts were being put into this tactic.

Air Defence Great Britain (ADGB) issued the basic 'Rules of Engagement' in mid-June:

> Aircraft, excepting four-engined bombers, burning navigation lights over the continent are to be attacked on sight. An aircraft without navigation lights might be attacked if it was acting in a hostile manner – e.g. taking-off or landing at an enemy airfield, or if it had definitely been recognised as hostile by visual means.

The C-in-C Bomber Command, Arthur Harris, had asked for 100 intruder/bomber support sorties on each night that a major raid was mounted. Such a requirement stretched resources to the limit, so it was decided to use a number of single-engined-fighter squadrons for 'night rhubarbs' (Nos. 1, 3, 198 and 609 being chosen) during moonlight periods.

Meanwhile, night after night, Bomber Command was taking the war to Germany. The following debrief report was provided by Sergeant O'Leary, the bomb aimer of Halifax DK239/Q of No. 428 Squadron, lost on the Aachen raid of 13/14 July 1943:

> About 10–15min after crossing the coast, Monica started to give a warning. The captain ordered the crew to keep a sharp lookout for enemy fighters and continued on a weaving course. In the act of taking a pinpoint, the navigator saw a stream of red tracer pass under the nose. Monica had been giving a continuous warning for 3–4min before the attack, but the pips did not appear to become more rapid and no-one reported seeing the fighter. It felt as if the Halifax had been hit by the first burst of fire and I fired a burst across the line of the enemy's tracer without seeing the fighter. Looking up, I saw that the emergency warning light was flashing – the prearranged signal to bale out. I turned around and saw the navigator crouching by the escape hatch with his parachute on . . . noticed the pilot having great difficulty holding the aircraft straight and level, the control column was pressed right back and he was applying port and starboard aileron alternately in rapid succession . . . I baled out.

O'Leary landed safely, evaded capture and escaped back to England. Three other crew members were killed and three taken prisoner. Of the 374 aircraft on the raid, 20 failed to return, including 15 of the 214 Halifaxes.

Monica, introduced into service in mid-June, was intended to give bomber crews warning of the presence of aircraft closing in from behind. A small transmitter installed in the tail of the bomber had a range of from 1,000 yards to 4 miles and should, in theory, have given the rear gunner advance notice of the fighter sneaking up from behind. The warning was an audio one; the nearer the contact, the more rapid the 'pips'. However, this proved to be a nuisance, as in the bomber stream the equipment was picking up almost constant contacts from other bombers. One attempted solution to this was the Mk. IIIA, with a visual indication (lights), though the problem of too many contacts remained. Worse was to come.

After months of debate, and in the face of an unacceptable loss rate, the use of one of the simplest but most effective countermeasures devices was eventually sanctioned. As early as 1941 the TRE had proved the concept of using aluminium strips dropped from aircraft to confuse a radar picture, but it was feared that if this 'window' was used, the Germans would copy the idea and use it against British radars. Agreement was reached only when Arthur Harris insisted, in July 1943, that the use of this material could cut his Command's losses by up to 30 per cent. For some time, losses on the major raids had been running at 4–5 per cent, a rate that would be insupportable in the long term. The Hamburg attack on the night of 24/25 July by 791 bombers was to be the operational trial for 'window'. It proved an immediate success, throwing the German defences into confusion. Only twelve bombers were lost (1.5 per cent).

Johnen recalled the problems caused by this simple device:

> It was obvious that no-one knew exactly where the enemy was or what his objective would be. An early recognition of the direction was essential so that the night fighters could be introduced as early as possible into the bomber stream. But the radio reports kept contradicting themselves. Now the enemy was over Amsterdam and then suddenly west of Brussels, and a moment later they were reported far out to sea in map square 25. The uncertainty of the ground stations was communicated to the crews . . . All the pilots were reporting pictures on their screens. I was no exception. At 15,000ft my sparker announced the first enemy machine on his Li [Lichtenstein]. Facius proceeded to report three or four pictures on his screen. I hoped that I would have enough ammunition to deal with them all! Then Facius suddenly shouted: 'Tommy flying towards us at a great speed . . . 2,000 yards . . . 1,500 . . . 1,000 . . . 500.' This was soon repeated a score of times. Then the ground station suddenly called: 'Hamburg, Hamburg, a thousand enemy bombers over Hamburg. Calling all night fighters, calling all night fighters, full speed for Hamburg.'

By the time he reached the area it was too late. The bombers were on their way home.

Meanwhile, in the Mediterranean theatre, plans were being laid for the next stage in the Allied advance, the invasion of Sicily. As part of Operation Husky, launched on 10 July 1943, consideration had been given to the problem of providing a nightfighter defence for the vulnerable beach-heads. Four nightfighter units, No. 600 Squadron with Beaufighters and Nos. 23 and 256 Squadrons and 215 Flight with Mosquitoes, plus flights of No. 108 Squadron (Beaufighters) and No. 73 Squadron (Hurricanes) had been allocated the task of night defence. All were initially to operate out of Malta using Malta GCI, but with the added benefit of having forward control posts, as three GCI units had been embarked on LSTs, one for each beach (Acid, Dime and Bark). It was said that it was 'a novel experiment and the results achieved clearly proved its future value'. Those results were indeed impressive:

10/11 June	Two Ju 88, one Cant 1007
11/12 June	One He 111, two Ju 88, one unidentified
12/13 June	Five Ju 88, three He 111, two Cant 1007, one Do 217
13/14 June	Two Ju 88, two He 111, one P.108, one unidentified
14/15 June	Nine Ju 88, two He 111, one Cant 1007
15/16 June	Six Cant 1007, five Ju 88, two He 111

In the meantime, in the face of growing problems for the Luftwaffe night defences resulting from a concerted initiative by the RAF to regain control of the night skies through the employment of RCM and intruder aircraft, the organisation of the night defence system was reviewed – and Kammhuber was replaced as General der Nachtjagd by Generalmajor Josef Schmid. It was considered that both Kammhuber and his 'line of defensive boxes' lacked the flexibility to deal with the ever-changing and growing bomber threat. Among the proposals that now came up for discussion was the reintroduction of night intruder operations over England. The idea was for large-scale employment, but confined to selective, favourable nights when the results would justify the risk. A four-phase operation was envisaged:

Phase 1	Bf 110s pursue the bombers on the return flight.
Phase 2	Me 410s of Fliegerkorps X (N France) carry out deception intruder missions into west and south-west England, to confuse the defenders.
Phase 3	Ju 88s patrol the British bomber bases when the first bombers return.
Phase 4	About 50min later another wave of Ju 88s arrives over the bases.

It was also suggested that the use of English-speaking crew in selected aircraft should be used to issue false orders.

Fortunately for Bomber Command the idea was rejected by Goering and many other senior commanders, partly on the grounds that Hitler was still very much against 'Fernnachtjagd' (long-range night fighting) because he wanted the destruction of bombers over the Reich. This did not, however, mean that limited intruder operations were forbidden, and from July onwards the Me 410s of V/KG2 undertook such missions. It was a period of mixed fortunes. After a 20min chase on 13/14 July, one of these aircraft was shot down by a No. 85 Squadron Mosquito, the *coup de grâce* being delivered by a two-second burst into the underside of the aircraft at 200 yards range. There were also successes for the Luftwaffe crews, such as that of the early hours of 24 August, when a No. 97 Squadron Lancaster was shot down over East Anglia by Oberleutnant Wilhelm Schmitte. However, he in turn was attacked by a nightfighter and his aircraft was damaged. Over the North Sea he was forced to bale out.

That same month, Fighter Command introduced a new offensive patrol, Mahmoud, whereby AI-equipped fighters were tasked against known German nightfighter assembly points. This was very much a 'cat-and-mouse' tactic, the fighter hoping to attract the attention of a prowling nightfighter. Hopefully, Monica would give an indication of the enemy presence; the trick was to wait until the range closed and then go into a tight turn to appear behind the other fighter, pick him up on AI or visually and make the kill. The major drawback to this scheme proved, once again, to be the poor performance of the Beaufighter.

At a conference on 23 August, Field Marshal Milch explained the problems, stressing that the day and night air attacks against German armament production had to be stopped or the war would be lost:

> Everything must be staked on the 110. Only the 110 in sufficient numbers can give us the necessary relief at night . . . Germany itself is the front line, and the mass of fighters must go for home defence. You can set up five times as many AA batteries; it will make no difference to the figure of 1–2 per cent (kills). But if we put twice as many fighters in the air, the number of successes will be at least twice as high. If we put four times as many fighters, the number of successes will be four times as high. [Such] success rates would make the enemy stop his bombing day and night.

Other senior commanders, such as Jeschonnek, did not see a problem:

> Every four-engined bomber the Western Allies build makes me happy, for we will bring them down just as we brought down the two-engined ones, and

> the destruction of a four-engined bomber constitutes a much greater loss to the enemy.

Many Allied leaders would have agreed. It now became a test of strength; could the bombers be destroyed in sufficient quantity, or would their bombing campaign prove decisive?

In recent months German fighter production had fallen by 25 per cent as a result of the Allied bombing, but the existing production dispersal plans were considered capable of arresting this situation if further strong attacks were prevented.

On the night of 17/18 August Bomber Command sent 596 bombers on Operation Hydra, the attack on the German research installation at Peenemünde. It was a moonlit night, but the plan included all the spoofing, deception and decoy tactics that the Command had been developing in recent months. The first two waves of bombers benefited from the plan, suffering only 11 losses, but the final wave of 166 aircraft was caught by the nightfighters that were racing to the area, and lost 29 aircraft. It is estimated that the diversionary missions caused some 200 nightfighters to move towards the wrong target, including a number that fell into what has been called a Serrate trap set up in the Friedland area by Beaufighters. Among those involved in the latter combats were five Bf 110s of IV/NJG1; they lost three aircraft. Total nightfighter losses amounted to 12 aircraft, nine having been shot down by bombers or fighters.

However, the most significant event was the probable first operational use by the German nightfighters of a new weapon, 'schräge Musik' (oblique or jazz music). Bomber Command ORS had been discussing for some time the possible employment by the enemy of an upward firing weapon, the evidence coming from crew reports and the nature of some of the damage suffered by aircraft that returned after being attacked by nightfighters. The system comprised a pair of cannon:

> The vertical guns were adjusted to slant slightly towards the nose. This gave the advantage of being able, when in a position reached by course and speed harmonisation, directly beneath the enemy aircraft, to fire in an upward direction. This had several advantages: firstly, the fighter was no longer within range of the quad guns of the tail gunner; secondly, it exploited the bomber's blind spot, where it had no view of the ground. Usually the bombers had no defensive weapons directly underneath. The dangerous aspect of this method was the possibility of being dragged down by the falling bomber. The art for the fighter pilot lay in hitting the bomber full in the underside and then manoeuvring himself smartly away. There was a further danger that the fighter's guns would hit the bomb load, in particular when

the bomber was carrying parachute mines, and there would be an explosion which inevitably would also destroy the fighter.

The system was designed by Paul Mahle, an armourer, after he had seen a similar concept used as a defensive system on a Do 217; the two 20mm MG 151 cannon were mounted on a hardboard panel as a Bf 110 fit.. Initial reactions among the pilots of II/NJG5 were rather sceptical, most preferring to use the deadly forward-firing armament of their aircraft. However, after the success of NJG5 against the bombers on Operation Hydra, Holker downing two and Oberleutnant Peter Erhardt five, opinions changed and the system received official backing. In its general principle the new system was almost identical to the experimental gun fittings tried by the RFC in 1917. Although Mahle is often credited as the 'inventor' of the system, this would appear, from German records, to be an overstatement in that from mid-1942 the weapon test centre at Tarnewitz had been experimenting with vertical weapon fits. Field tests had been conducted by 3/NJG3 in early 1943.

Most nightfighter pilots using 'schräge Musik' preferred to fire at the inner part of the wings in the hope of destroying fuel tanks and starting a fire that would destroy the bomber. Unless the first burst was badly aimed, there was little hope of escape for the bomber.

August was also a record month for intruder operations by the RAF, with 382 sorties, plus a further 101 Serrate and 12 Mahmoud operations (and 56 rangers by other aircraft, including Mustangs and Typhoons). Twenty-three RAF units had been involved during the month, claiming twelve enemy aircraft destroyed.

So far, few references have been made to operations outside the German Reich and Western Front. This has been intentional, as it was here that the major night war was played out, and that the majority of innovations were introduced. It has often been stated that the air war on the Eastern Front was a very one-sided affair, and to a large extent this is a realistic appraisal up to 1944. However, in common with the ground force, the Russian air force refused to give up. In an endeavour to gain some influence on the battlefield, the Russian bombers turned to night operations. To counter such raids, and the bombing of cities in Germany, the nightfighter system on the Eastern Front required major reorganisation. The Eastern Front had certainly been a 'make do' area. A number of day fighter units had done well: for example, Leutnant Leykauf of JG54 had shot down six Russian bombers on 23 June. On another night Unteroffizier Döring of 9/KG55 had brought down three TB-3s near Stalingrad using an He 111 (!), and later in the year even a Ju 87D-5 was brought into action by NJG100 to hunt down the slow and annoying PO-2s. However, overall success rates were less than 1 per cent.

The so-called 'railway nightfighters' (Eisenbahn-Nachtjagd) were controlled by GCIs operating from trains; hence the name. The main units were NJG100 and NJG200, primarily tasked with countering these nuisance raids. However, the only real solution was to transfer units, or, at the very least, experienced crews, from the western defences. Among those sent to improve matters in the East was Prince Wittgenstein, one of the leading nightfighter 'experten', who had already established a reputation over the western part of Germany.

During Operation Zitadelle (the Kursk offensive) a communiqué of 25 July stated: 'Last night Prinz zu sayn-Wittgenstein and his crew successfully shot down seven Russian aircraft. This is, to date, the highest number shot down in a single night'. This mission was recorded by the radio operator, Herbert Kummeritz, and repeated in a recent history of Prince Wittgenstein (*Laurels for Prinz Wittgenstein*, Werner Roell, Independent Books, 1994, p.103):

> . . . two 'Kuriers' who, about to bomb the German sector of the front, are on the point of crossing our path. The guns were already reloaded, so it required only a brief course correction from the wireless operator to bring the fighter into its firing position. And again the scene concentrated itself into an ominous complexity; searchlight beams seeking the aircraft, anti-aircraft shells exploding, hunters and prey equally threatening . . . both Russian bombers were smouldering on the ground . . .

Soviet night bombers also attacked important rear area targets such as the bridge at Temryuk and supply depots at Taman and Kerch. Germany's Crimea air assets were therefore reinforced by the arrival of a nightfighter detachment in July. In the Kerch area an effective tactic was developed:

> The heavy batteries opened fire the moment that searchlights spotted a plane, continuing to fire until the alerted night fighters signalled that they were ready to attack. The flak batteries then immediately ceased fire and the fighters attacked. By close co-operation it was possible to reduce the time interval between the batteries ceasing fire and the opening of the night fighter attack to less than a minute. Using these methods, by mid-July the Luftwaffe had shot down 59 aircraft, 20 being claimed by night fighters. [USAF Study No. 155]

Such improvisation was the norm for the Luftwaffe in Russia, having to cover such an enormous geographical area with inadequate forces. The absence of adequate numbers of dedicated nightfighters meant that a variety of types were employed; the Fw 58, Fw 189 and even the Ju 87.

The defences of East Prussia had to be reinforced continually from mid-1943 as the Russian bombers increased their nightly attacks on German territory. Elements of the Schleissheim training school were incorporated into NJG101 (Ingoldstadt) and NJG102 (Kitzingen), and by September each unit consisted of three Gruppen. Meanwhile IV/NJG6 moved to new bases to cover the Romanian oilfields.

Late 1943 saw the Luftwaffe introduce two new devices in the continuing electronics battle: Naxos and Flensburg. The most significant of these was the FuG 350 Naxos Z, designed to pick up the transmissions from the bomber's H2S radar. This could detect transmissions at ranges up to 30 miles, and gave the nightfighter a virtually foolproof means of locating the bombers. As H2S was still in short supply, there was a tendency to fit it to the pathfinder aircraft; prime targets for the fighters.

Now that Bomber Command operated in compact masses, the aim of the nightfighter pilot was to achieve 'mitschwimmend' (swimming with the stream). Having found the bomber stream, the pilot would stay with it and pick off targets in quick succession until his ammunition was spent. It was under such circumstances that crews achieved multiple victories, including the not-infrequent five-in-one-night scores. The new devices would certainly help this tactic. Flensburg was designed to home on to the transmissions from the Monica tail-warning radar fitted to all of the RAF's heavy bombers. Thus a device designed to protect the bomber against the fighter now became lethal.

Throughout the year the RAF had also been introducing new RCM devices aimed at reducing the effectiveness of the German night defences. Among those entering service during 1943 were Ground Grocer (April), to jam AI, and three systems designed to jam or confuse R/T control frequencies: Ground Cigar (July), Airborne Cigar (ABC, October) and Corona (October). The last of these went into use on 22/23 October. Bomber Command's major attack that night was against Kassel, and as soon as the German nightfighters were airborne they started receiving conflicting R/T instructions, thanks to German-speaking operators at the ground station in England. According the War Diary of 1 Jagdkorps: 'This was the first enemy use of spoof R/T orders to our nightfighters, but they were easily identified because of pronunciation'.

Nevertheless, it caused confusion, and over the next few months attempts were made to eliminate the problem. It was suspected that the

British interference was coming from aircraft (this was partly true, as Airborne Cigar was so used), and that the use of female controllers would solve the problem, as the English would not have females in their aircraft. This plan backfired, as it was found that many of the spoof females were better than their male counterparts!

The sheer scale of Bomber Command operations, with many hundreds of bombers on each mission, had caused a major re-think in the doctrine of the Reich defences. The adaptability of the German military throughout the latter years of the war deserves much praise. No sooner did the RAF come up with a new tactic to swamp or confuse the defenders than the latter came up with a counter, or at least a method by which they could still operate effectively. With streams of enemy bombers heading for a single point of the Reich, all that was required of the ground control was a running commentary of where the enemy were and where they were going. It was then up to the fighters, be they radar or visual, to make contacts. Typical of such commentaries was that for the night of 23/24 August:

2133	Bombers approaching Amsterdam
2155	Bombers flying east; orbit searchlight beacons
2217	Bombers approaching Bremen
2238	Berlin is possible target
2304	All fighters proceed to Berlin
2332	Bombers over Berlin

In this way the controller could keep his options open, not committing his forces too early and ensuring that he had the maximum forces to counter the main attack. The fighters could move from beacon to beacon, keeping pace with the stream until the time to attack. As soon as a fighter found the bomber stream, often with the aid of the commentary from 'observer aircraft' that, having found the stream, joined it and passed position updates, it was down to individual combat techniques. It was also common for fighters to drop flares or markers around the stream to attract the attention of their colleagues. To help counter such actions the RAF flew sweep and escort AI nightfighters. Once in the bomber stream, a fighter using 'schräge Musik' could easily make multiple kills, moving from one radar or visual contact to another. The stream was so concentrated that there was a plethora of targets.

There were 727 aircraft in the 23/24 August attack against Berlin, and the fighters laid claim to at least 33 of the 56 aircraft lost that night, more than half of those being over and around Berlin. Losses to the fighters might well have been higher but for the fact that they were recalled just before midnight because fog threatened to blanket their bases.

The story was similar on subsequent missions to Berlin, with loss rates far in excess of the sustainable 4 per cent. The defenders appeared to be winning the battle in the night skies over Germany. What the British did not know was that the defenders were having their own problems, not least of which was the high accident rate among the nightfighter units, all of which combined to reduce the availability of aircraft and crews.

While the 'Wilde Sau' fighters were less prone to interference from the RAF's bag of electronic tricks, there was a growing feeling that the single-engined fighters were now of little value. On 29 September Milch stated:

> Wilde Sau operations no longer have any prospect of increasing the number of kills over a target beyond a specific maximum. Himmelbett operations must be at least tentatively reintroduced and stepped up because the enemy have recently been flying in looser formation on a fairly broad front. There has also been a sharp drop in the serviceability of night fighter formations, and a first priority must be to maintain serviceability for the coming difficult winter of 1943/44; furthermore, the autumn weather conditions will increase the difficulty of Wilde Sau operations.

The Operations Diary of 1 Jagdkorps records that Wilde Sau operations were terminated on 5 October 1943, but this does not appear to have been the case, as in subsequent raid reports such missions still comprise the majority of sorties, though the single-engined types appear to have vanished. Typical of the operational ratios for this period was that of 2/3 October, with 193 nightfighters operating against a Bomber Command raid (by 294 Lancasters) aimed at Munich. The War Diary records:

> All XII Fliegerkorps units employed on Wilde Sau were brought into the Augsburg-Munich area well before the attack commenced, but dense haze and absence of moonlight prevented any great measure of success. Moreover, the enemy is believed to have made considerable alteration in his altitude just before and during the attack, with the objective of impeding night fighter interceptions. Only 21 contacts were made.

The night had seen 87 Bf 110s, 25 Ju 88s and two Do 17s on Wilde Sau, and 28 Bf 110s, 38 Ju 88s and 4 Do 17s on Zahme Sau ('Tame Boar'). The nightfighters claimed ten bombers for the loss of three of their number; RAF records show a loss of eight Lancasters to all causes on this raid.

Royal Air Force countermeasures had greatly reduced the effectiveness of Lichtenstein B/C, so the introduction of a new AI radar was an urgent requirement. At last, in October, the SN2 system became operational.

Operating between 37.5 and 118 mHz, the new AI was, at first, virtually immune to the RAF jammers. The maximum range was just over 3 miles at approx 18,000ft, but the minimum range was just under 1,000ft. The latter proved problematical, as crews lost their targets in the critical final stages. The adopted solution was to make use of both AI systems, the operator reverting to the old Lichtenstein B/C for the final stages of the intercept. The combination of both sets of aerials on the aircraft's nose caused a substantial reduction in its speed, though this was still sufficient to enable the pilot to stay with the slow bomber stream.

In October 1943 Bomber Command ORS issued a series of notes on tactics, dealing with the basic organisation and operation of the German defensive system. In part, these stated:

> The German Air Force uses GCI. This starts when coastal Freyas of the Aircraft Reporting Service warn Sector HQs of approaching aircraft. Fighters get airborne to orbit forward beacons of the box to which they are allocated, with maybe a reserve fighter at an inner beacon. The GCI follows the fighter on a Giant Würzburg, when a raider enters the box it is tracked by a second Giant Würzburg and the fighter homed by R/T. The fighter takes over when it has visual or AI contact with the enemy. Post-combat the nightfighter returns to orbit the beacon . . . The aim of the night bomber is to gain immunity from attack through concealment. Make the GCI's task harder by weaving, including use of height – this is best at short range, say 1–2 miles, just before AI contact – then put in an orbit to throw off the fighter. However, this will only be possible when the radio-location aids are available. Until then it is best to go straight and level, using the concentration corridors.
>
> Over the target fly straight and level unless picked up by searchlights, keep above 15,000ft to avoid the worst of the flak. Best to get out of gun-defended zones as soon as possible. It cannot be too strongly emphasised that when massed searchlights fasten on to an aircraft, which happens swiftly, and is followed almost immediately by the noise and buffet of continuous near shell bursts, no human being can consider calmly the best way to turn. Escaping the searchlight cone is more a matter of mental discipline than tactics.
>
> *Air combat.* The endeavour of the bomber is to finish the combat without being seriously damaged, and this can only be achieved by manoeuvring so that the fighter's fire goes wide, and making up for the lighter armament of the turret by firing accurately at the fighter. During the corkscrew the pilot should call the next manoeuvre so that the gunners can allow for it . . . 'going down, port' . . . 'going up, starboard' . . . the striking power of a .303 is adequate to damage the frontal portion of a nightfighter, which has a large frontal area scarcely protected by armour.

October brought another major headache for the Reich defences with the inauguration of major raids from bases in Italy, the Allied acquisition of air bases around Foggia having added a new dimension to the Combined Bomber Offensive. The Reich was now being assailed from three sides, day and night, and the fact that the defences did not simply collapse testifies to their adaptability and determination.

The continuing high attrition rate of nightfighters was causing its own problems, as the following memo to the Personnel Office of the Reichs Air Ministry, dated 17 October, illustrates:

> In recent nights, heavy casualties have occurred among the commanders of Gruppen and Staffeln. If these losses continue, suitable Gruppen commanders would not be able to be found from the ranks of the night fighter formations. It is requested that about ten commanders of bomber units with operational and night flying experience be transferred immediately.

Other memos and notes comment upon the lack of training and experience of many nightfighter crews. There was a similar strain on ground personnel, especially in the radar and GCI units. With the increased availability of SN2-equipped aircraft, it was considered possible to reduce the number of nightfighter Areas from 39 to 19, with a consequent significant saving in personnel. This, added to other improvements, such as the widespread installation of 'Gemse-Erstling' (an IFF device), was intended to streamline and improve the Zahme Sau technique while also allowing flexibility for free-ranging missions. In view of the almost inevitable heavy jamming that accompanied every bomber raid, one of the most useful elements had proved to be the 'observer' aircraft. The basic concept was simple. The aircraft would locate, and frequently join, the bomber stream, but, rather than making any attacks, would relay position information to the Air Reporting Service.

The nightfighter defences had been strengthened throughout the year, and the same was true of the other elements of the air defence network. At the beginning of 1943 the Luftwaffe's anti-aircraft strength in the west stood at 5,421 heavy guns, 9,528 light/medium guns and 3,726 searchlights. By the following January these figures had risen to 7,941 heavy guns, 12,684 light/medium guns and 6,880 searchlights; a major increase. (These figure relate only to Luftwaffe strength. Army and Navy units are not included.)

The burgeoning radio countermeasures organisation within the RAF was reorganised on 23 November with the creation of 100 (Special Duties) Group, whose duties were to:

1. Give direct support to night bombing or other operations by attacks on enemy nightfighter aircraft in the air or by attacks on ground installations.

2. Employ airborne and ground RCM equipment to deceive or jam enemy radio navigation aids, enemy radar systems and certain wireless signals.
3. Examine all intelligence on the offensive and defensive radar, radio and signalling systems of the enemy, with a view to future action within the scope of the above.
4. Provide immediate information, additional to normal intelligence, as to the movements and employment of enemy fighter aircraft to enable the tactics of the bomber force to be modified to meet any changes.

The 'teeth' for the offensive task were provided by Mosquito squadrons transferred from other Groups, the first being No. 141 Squadron, who moved to West Raynham, Norfolk, in early December. By mid-December the Group had four such squadrons plus two special flights. Although, as related above, Serrate had already been used, its operational outing with 100 Group came on the night of 16/17 December. Bomber Command mounted a major raid of 483 Lancasters (plus 10 Mosquitoes) to Berlin, accompanied by two Mosquitoes and two Beaufighters of No. 141 Squadron. A number of contacts were made, and the Mosquito of Squadron Leader Freddie Lambert and Flying Officer K. Dear managed to engage a Bf 110, claiming it as damaged.

Flying Officer M. Kelsey and Pilot Officer E. Smith of No. 141 Squadron were airborne in Beaufighter VI V8744 on one such patrol in late December. Their combat report is typical of such missions:

> Landfall made at Overflakee 0118hr at 24,000ft. There was 10/10 cloud over Holland but the cloud was breaking in the Ruhr and to the East. Haze persisted up to 20,000ft and visibility was moderate. When 25 miles inland and until SW of the Ruhr between 0125–0155 flying at 24,000ft, numerous weak Serrate contacts were obtained. They came on and off in pairs to port and below, and appeared to be at least 20 miles, and possibly as much as 50 miles to the North. About 10 miles NE of Aachen at 0155 there were two more Serrate contacts hard starboard 10 miles range. As they were 25 above and aircraft was already at 20,000ft they were not followed. A/C patrolled Uckerath from 0200–0205 but there was no sign of activity and nothing could be seen on the ground through the clouds. On the way back to the Duren area, a Serrate contact was obtained 50 starboard and 5 below and 10 miles to the north. A/C turned at 300 ASI towards E/A who was flying south very fast, and decreased height gently to get below the E/A. When 5 miles away E/A turned due east, turned with it to reduce the distance and shortly afterwards AI contact was obtained at 14,000ft range. When immediately behind and slightly below, visual was obtained on E/A which was a Ju 88 burning green and white resins. Pilot was of the impression that the E/A was drawing away so opened fire at 1,000ft range, with cannon and MG, causing

E/A to slow down. Pilot continued his burst for 7 seconds closing to 300ft. Strikes were seen all over E/A who turned steeply to port and dived straight down with flames streaming from both engines and the fuselage, to explode on the ground where a glow could be seen through the clouds. This E/A is claimed as destroyed. 630 rounds of .303 and 246 rounds of 20mm were used. Immediately after the combat over Aachen, A/C was hit by intense and accurate heavy flak. The port engine was running roughly and various parts of the leading edges and wings were damaged. A/C set course for home. Serrate contact obtained 5 starboard and 5 below. A/C turned head-on into the E/A which passed 300ft starboard and gave chase from behind. Serrate contact remained until shaken off in the Moll area.

November also brought a long-awaited victory for the defenders of the Reich, with the destruction of a high-flying Mosquito by a special Ju 88R-2 equipped with a nitrous oxide auxiliary fuel system. This much-vaunted victory does not appear in RAF records. However, at about the same time the RAF, too, were experimenting with nitrous oxide as a way of providing increased performance above 20,000ft. Trials were conducted by the FIU and No. 85 Squadron, and the early results looked promising, the nitrous-oxide injection providing a boost of around 47mph; the aircraft carried enough for six minutes of use. The first operational success came early the following year, when John Cunningham in Mosquito HK374 of No. 85 Squadron shot down an Me 410 near Le Touquet on 2/3 January. As a result, it was intended to have 50 Mosquito NF.XIIIs modified by Heston Aircraft for use by Nos. 96 and 410 Squadrons.

Meanwhile, German attacks on the UK had continued, still small-scale but nonetheless threatening. On one night, 5 November, Bud Green was airborne in a No. 410 Squadron Mosquito:

I was flying a patrol line in the middle of the Channel, with other Mosquitoes flying on my right and on my left. Ground control could only handle a limited number of aircraft; it was sited at Wartling near Dungeness. I was instructed: 'Bandits on a southerly heading, height unknown, considerable numbers.' We were taken over by a controller who steered us towards them, then round them (at 18,000ft), until we picked them up on our own radar. A classic interception.

Once the navigator had picked up an aircraft on his AI, the ideal was for the pilot to come in below at a closing speed as slow as possible, but fast enough for evasive action if fired on. You had to match the speed before you got a visual contact, then open up the throttles to close in by eye. I came in too low, overtaking speed was not high enough. I had to pull up at full throttle, but was still too slow. Pointing upwards at 200 yards, I fired all four cannon and the four machine-guns. We were so close, I didn't need tracer or to allow for any deflection. He never knew what hit him. It was totally devastating as he just blew up. I thought it was an Me 410 but it was a Do 217.

The other major change for the RAF at the end of the year was a reorganisation of squadrons, with ADGB strength being reduced to ten-and-a-half nightfighter squadrons, and the remainder going to 85 Group as part of the Allied Expeditionary Air Force (AEAF); 1944 was to be the year in which the Allies returned to the Continent.

In 1943 technology had become increasingly important to both attacker and defender, as first one and then the other introduced a new item of electronic equipment in an effort to achieve a tactical advantage. Bomber Command had increased the weight and effectiveness of its attacks once more, and in response the German nightfighter force had been built up towards what would soon be its peak efficiency. On some nights the bomber losses had been horrendous as the nightfighters entered the bomber stream and wrought havoc. However, the nightfighter force had also suffered grievous losses, including the deaths of a number of 'experten' crews. The battle for the night skies over Europe was still evenly poised.

As with the Russian Front, it is not intended to cover Pacific operations in any detail, partly because the basic nightfighter lessons and problems were universal. However, the main reason is that night operations in this theatre of war were invariably small scale. United States Army Air Force statistics show that there were only 51 nightfighter crews in the Far East and seven in the Pacific in the period March to December 1943; however, these figures do not take the naval air elements into consideration.

The 6th Night Fighter Squadron (NFS) had formed out of the 6th Fighter Squadron on 9 January 1943 and, as part of the 15th Fighter Group (FG), deployed a detachment (Detachment B, under Major Sydney Wharton) to Guadalcanal in February. The P-70 conversion of the Douglas A-20 light-bomber very much followed the earlier British example (*à la* Havoc), but was intended as a stop-gap until the definitive Northrop P-61 became available. America was very late in entering the nightfighter 'market', not having seen any great need for this type of aircraft. The P-70 was given the British AI Mk. IV, a crew of two and a primary armament of four nose-mounted 20mm cannon. Its only operational service was to be with the 6th NFS in the Pacific, but it also played an invaluable role as an AI trainer with almost every nightfighter squadron, as well as with the 481st Night Fighter Operational Training Group. It was produced in a number of variants, depending upon the A-20 parent version and the mark of AI installed (later ones having the SCR.720 set).

The Japanese Mitsubishi G4M 'Betty' bombers were tasked with carrying out night nuisance raids on Guadalcanal, where they acquired the nickname 'Washing Machine Charlie' because of the noise they made. The anti-aircraft guns proved ineffective, so pressure mounted for some kind of air defence, the suggestion being to employ a Lockheed P-38/searchlight combination. However, the interim nightfighters did not usually have the performance to reach the bombers, although the first nightfighter kill was made on 19/20 April by Captain Earl Bennett and Corporal Edwin Tomlinson, the victim being a 'Betty'.

To return to the P-38, a detachment from the 347th FG joined the 6th Night Fighter Squadron. Among these new pilots was Lieutenant Henry Meigs. On the night of 15 August he downed a 'Betty', having used the basic technique of watching the searchlights cone the aircraft, keeping fingers crossed that the AA guns left him alone, and closing in and shooting down the bomber. A few weeks later, on 20/21 September, a force of six G4Ms from the 702nd Kokutai left Rabaul to attack one of the Guadalcanal airfields. Henry Meigs was scrambled:

> By the time I reached altitude and turned inland the searchlights were on the first Betty and all our 90mm guns were filling the sky with brilliant flashes. As I closed on this Betty I tried to fly directly astern but keeping back far enough to stay out of the searchlight aura. At about 200ft I fired the four .5in guns at the right wing root. Before I could squeeze the trigger on the 20mm cannon, the Betty burst into flame and nosed over in a steep dive toward the ground.

He was quickly directed on to another bomber which he also shot down, the whole engagement from first sighting to second victory taking less that a minute.

Lieutenant Fred Secord served with the 6th NFS:

> The basic P-70 was armed with four 20mm cannon in a 'gun tub' attached to the bottom of the fuselage, where the bomb bay doors would have been on an A-20. On each side of the fuselage was a single .5in machine-gun, these being loaded with tracers. In theory, the machine-guns would be used first as an aid to aiming. On impact the shells would splatter, giving the pilot visual indication of being on target. At that time the 20mm cannon would be fired. All the guns were bore-sighted to converge at about 600ft.

In March the 6th NFS sent Detachment A to New Guinea, six P-70s operating out of Port Moresby. The air defence network was still in its infancy, but the basic system relied on a radar pick-up to scramble the nightfighters. Only one success was achieved during this period, Lieuten-

NAKAJIMA J1N1 GEKKO (MOONLIGHT)

The original requirement that Nakajima attempted to fill was for a twin-engined long-range fighter to a specification issued by the Japanese Navy in 1938. Although the aircraft first flew in October 1941, it was subject to a great many teething troubles. Falling well short of the requirement, it was rejected as a fighter but accepted, if redesigned, as a long-range reconnaissance machine. As such it entered service as the J1N1-C ('Irving') in July 1942, seeing action in the Solomons campaign.

Among the units so equipped was the 251st Kokutai at Rabaul in Papua-New Guinea, and this unit decided to make a few unofficial modifications in an attempt to find a solution to the need for a nightfighter. The main change was to add slanting cannon above and below the fuselage. Success soon followed, and the idea was put forward for official sanction. The Navy liked it, and ordered Nakajima to work on a true nightfighter. Between August 1943 and March 1944 some 183 Gekko nightfighters were produced, and a further 240 came off the production lines by the end of the year, when production was terminated. The type proved reasonably successful against the slower bombers, but was outclassed when the B-29 appeared. A number of aircraft were equipped with nose searchlights as an aid to interception. The biggest error was the use of the aircraft by day; valuable aircraft and crew were simply wasted.

Gekko data

Powerplant: Two Nakajima NK1F Sakae. Crew: Two. Wingspan: 55ft 8½in. Length: 39ft 11½in. Maximum speed: 329mph at 18,000ft. Ceiling: 34,000ft. Armament: Various combinations of cannon.

ant Burnell Adams and Flying Officer Paul DiLabbio shooting down a Mitsubishi Ki-21 'Sally'. DiLabbio describes the interception:

> I picked up the first bogey at approx 18,000ft range and quite lower than our altitude of 26,000ft. I gave Lt Adams directions to dive and turn port. At this time our intercom began to play up and I was only able to give one more set of instructions, which were to level out of the dive and steady the heading. Immediately, Adams said he had a visual and seconds later I heard the 20mm being fired. Adams scored several hits and the Japanese bomber exploded. While Adams was firing, I picked up the second bogey and had him pinpointed, but the intercom was garbled and I could not pass the information.

Japan had seen even less need for a true nightfighter, the entire doctrine being one of offence, in even more extreme fashion than that propounded by Luftwaffe doctrine. Although the standard day fighters, such as the Mitsubishi Zero, were occasionally employed at night, it was invariably with no success but at substantial hazard to aircraft and pilot. The 4th Air Group at Rabaul had used the Zero-searchlight combination since January. For finding the enemy, total reliance was placed on searchlights and weather conditions (bright moonlight). There was not even a rudimentary air defence system. However, the first 'true' nightfighter was brought into service in 1943. This was the Nakajima J1N1 Gekko (Moonlight) series, known by the Allied codename 'Irving'. The type had originally entered service as a long-range escort fighter but, just like the Bf 110, had proved defective in this role. Initial development work was conducted by Air Group 251 at Rabaul, under Commander Yasuna Kozono. Intercept techniques and weapon improvements were proposed, the latter including the installation of upward-firing cannon. The first operations with the new nightfighter equipment took place in May 1943. On the night of 21 May, Chief Petty Officer Kudo was on patrol looking for B-24s. The bright moonlight helped, and he soon spotted his quarry, shooting down two of the bombers. By the end of June he had claimed a further six bombers. The Naval Staff approved the modification to the aircraft and instructed Nakajima to produce a true nightfighter variant, work commencing in August. As so often proved to be the case during the Second World War, innovation was in large measure down to individuals or units, who then had to 'persuade' higher authorities to pursue the matter.

Chapter 7

THE PEAK OF LUFTWAFFE NIGHTFIGHTER EFFICIENCY

The situation at the turn of the year was such that many in Bomber Command were expressing concern at the continued high loss rates, attributed primarily to the Luftwaffe nightfighter force, although flak was by no means an insignificant element. In January the total of Luftwaffe flak units covering Germany and the Western Front had increased to 7,941 heavy guns and 12,684 light/medium guns (along with 6,880 searchlights). This total did not include army and navy flak units. Conversely, the nightfighter force had shrunk, albeit marginally, with a strength of just under 600 aircraft and a daily availability of about 400. Despite the German success, the picture for the defenders was by no means simple. Losses were running at around 3 per cent, deliveries of equipment were slow, serviceablility was poor and the electronic war was proving increasingly difficult – and, furthermore, Bomber Command was continuing to destroy German cities.

The contribution made by the flak arm was considered vital, and frequent co-ordination meetings were held; 'the prime objective must always be to bring that arm into action which would prove most effective at any given time'.

Throughout 1943, and in the face of what might have appeared to be overwhelming problems, the Luftwaffe air defence system had developed into an effective force. The most important single factor was the extensive development and integration of the early warning system. The nightfighter arm had also been forced to become flexible in all that it did, capable of rapid adaptation to whatever countermeasures the enemy might devise. The air battle was poised to go either way, and both sides knew that the winter of 1944 was going to be crucial.

By January 1944 the so-called Battle of Berlin was at its height. However, the Main Force target for the night of 21st/22nd was Magdeburg, with 648 aircraft (421 Lancasters, 224 Halifaxes and three Mosquito target markers) taking part. The German nightfighter control system was slow to identify the target, possibly assuming that Berlin would turn out to be the aiming point, as it had been on so many nights in the previous two months. Nevertheless, some Luftwaffe fighters had already found the bomber stream by the time it approached the German coast, and losses began to mount among the raiders. One of the nightfighters was flown by a man who had already been acknowledged as the leading exponent of the art. The most significant event this night was to be the death of Major Heinrich von Wittgenstein.

During the first three weeks of January von Wittgenstein was credited with shooting down 15 bombers, taking his total to 83, ahead of Helmut Lent's score. The story of this fateful night was told by his wireless operator/navigator, Feldwebel Friedrich Ostheimer:

> According to reports from ground control, the bomber stream was flying was at 8,000m [25,000ft]. The first anti-aircraft fire could be seen, so the bombers could not be very far away. In the meantime, we had reached the same altitude as the enemy aircraft. Like a signal to attack, an aircraft burst into flames in front of us and went down like a burning torch. Now our pilot made a right turn on to the same course as the bombers. The hunt in the dark began again. For some time we had been not only hunters but also the hunted. Ever more British long-range nightfighters accompanied the bombers to hold losses within limits. On my radar the first target showed up. After a small course correction we had it right in front of us, flying rather higher than our machine. The distance shrank visibly, so it was most likely a bomber. As the shadow of the enemy aircraft pushed slowly over us, it could be recognised as a Lancaster against the starry sky. After only one burst from the 'schräge Musik', the left wing was immediately in flames. The burning Lancaster went down into a shallow dive, which then turned into a spin. With an enormous explosion the fully laden bomber crashed, its lethal cargo exploding with it. It was between 2200 and 2205hr.
>
> We went on searching and at times I had up to six different targets on my radar screen. We had quickly carried out two course corrections and already had the next target in front of us, another Lancaster. A burst of fire and, to start with, the aircraft burned a little, then dived steeply over the left wing. Shortly after that I saw the fire as it crashed. It was about 2220hr. After a short run-up another target had been found and had been knocked down with a heavy burst. It must have been around 2230 when I noted the crash. Then everything went very fast indeed. A few minutes later we had another four-engined bomber in our sights. Like a pike in a lake full of carp we were now flying in the middle of the bomber stream.
>
> This enemy bomber too went down in flames after the first attack. It was 2240hr when I observed the crash. I already had the next target on the radar screen. After two small course corrections we had another Lancaster in front of us. After a single attack the fuselage was alight, but the fire became smaller and smaller and we approached for another attack. We were soon in position

> and Major Wittgenstein was about to fire when there was a banging and flashing in our machine. The left wing immediately caught fire and the aircraft began to dive. Then I saw the cabin roof above our heads fly away and heard a shout on the intercom which sounded like 'Raus!' [Out!]. I tore my oxygen mask and helmet off and was catapulted out of the machine. After a while I pulled the parachute release and landed about 15min later near Schönhausen. Insofar as I could reconstruct what happened, we had been fired at from below. I had looked around but could not see another parachute.

Ostheimer had come to earth at around 2320. The flight engineer, Kurt Matzuleit, also made a successful parachute descent, but Wittgenstein's body was found near the wreck of his aircraft. The subsequent inquiry concluded that he had been rendered unconscious through hitting part of the aircraft, and so was unable to deploy his parachute.

At a Divisional Commanders' conference on 25 January, Schmid stated that:

> . . . the number of aircraft being shot down by day and night is still too low and that better results should be possible, given full exploitation of all resources. Zame Sau had proved itself to be the best nightfighter operational technique and such aircraft should be fed into the enemy stream carefully and at an early stage.

The following day an order was issued to disband one Staffel within each nightfighter Gruppe. This was followed by a series of conferences at the highest levels of the Luftwaffe command, culminating in a major reorganisation of the Reich air defences into three areas of responsibility:

1 Jagdkorps	Hanover–Magdeburg
2 Jagdkorps	Oldenburg–Bremen–Rothenburg
3 Jagdkorps	German–Dutch frontier

Experienced crews were being lost, new crews were given inadequate training, and tired crews were not allowed enough rest; the strain on the squadrons was becoming extreme. Considering the difficulties, they were achieving major successes: it was not beyond reason that, given adequate resources, they could drive Bomber Command from the night skies.

Following the recent heavy attacks on Berlin, Hitler was determined that reprisal attacks should be made against London. The night of 21/22 January brought Operation Steinbock, the Luftwaffe's largest raid for some time. The attack comprised 447 sorties in two main attacking waves, primarily of Ju 88s and Do 217s. The bombers used 'duppel' (window) in an attempt to blot out the GCI network, but in the event many of the attackers did not even cross the English coast, and of those that did not many actually bombed their targets in London. Nevertheless, it was still the busiest night that the RAF defenders had seen for a while, and most squadrons were active at some time. Among those scoring victories was Flight Lieutenant J. Hall of No. 96 Squadron, who claimed a Do 217 and a Ju 88, the latter subsequently being identified as B3+AP of 6/KG54. Another attacker that failed to make it home was a Heinkel He 177 of 1/KG40, the first of the type to be shot down over the UK (this was the first time they had been employed). It fell to Warrant Officer Kemp of No. 151 Squadron.

According to the RAF narrative:

> The defences were seriously handicapped, firstly because of the enemy's extensive use of window, secondly because friendly bombers were returning from northern France. They gave, nevertheless, a good account of themselves. Air Defence of Great Britain flew 96 sorties and together with the anti-aircraft guns claimed to have destroyed 16 enemy bombers. Searchlights were invaluable in assisting fighters to shoot down several of the enemy. In fact German operational losses through Allied action amounted to 25 bombers, plus 18 lost to other causes. In none of the subsequent raids on London were so many aircraft despatched in one night, nor were such heavy losses experienced.

The window comment is interesting, as in previous reports this was claimed to have had little effect. At this time ADGB comprised eighteen squadrons: five with Mosquito XII/XIIIs (with AI Mk. VIII), two with Mosquito XII/XVIIIs (with AI Mk. X), five with Beaufighter VIs (with AI Mk. VIII), four with Mosquito IIs (with AI Mk. V) and two intruder squadrons with Mosquito VIs. Although 11 Group's squadrons were the busiest, the other Groups were drawn into the battle whenever the Luftwaffe raiders extended the area of their attacks.

The raids during February produced mixed results, but included a number of highly accurate attacks, particularly those of the 18/19th and 20th/21st, although losses, especially among the Ju 88 formations, remained high. This so-called 'baby Blitz' petered out during March. During this period the defenders had seen a marked improvement in their performance owing to the widespread introduction of AI Mk. VIII and, in June, the superior AI Mk. X. By the end of March, aircraft equipped with the latter system had claimed 23 enemy bombers. AI Mk. X was a development of the American SCR.720B, and although the first sets had arrived in the UK as early as December 1942, progress with this ECM-capable set had been slow. The first unit to re-equip was No. 85 Squadron, in January 1944, which flew the first operation on the 12th and scored the first victory (a double) on 20 February.

The nightfighters still had the problem of dealing with high-speed bombers such as the Me 410. Even the Mosquitoes were hard-pressed, so a stop-gap was needed pending the introduction of the Mosquito XXX. The solution adopted was to use the N_2O (nitrous oxide) injection system tested by No. 85 Squadron. Air Chief Marshal Leigh-Mallory ordered 50 modification kits to go to Nos. 96 and 410 Squadrons for the AI Mk. VIII-equipped Mosquito XIIIs; most were installed by the end of March.

Following the failure of the attacks on London, and in the face of mounting losses, the Luftwaffe bomber force turned its attention to less well defended targets such as Hull and Norwich. However, the defences were up to the challenge and, on the night of 19/20 March, for example, Flight Lieutenant Singleton in Mosquito HK255 of No. 25 Squadron shot down three Ju 188s that had been involved in such a raid. In the three months ending 18/19 April the Luftwaffe had lost 139 bombers to British defences, plus another 59 to other causes, approximately 6 per cent of those despatched. There was also a high accident rate through lack of crew experience.

Meanwhile, in the Mediterranean, the Allies were renewing their offensive in Italy, having broken through the German winter defence line in early January. Air power remained a vital element of the campaign, so the nightfighter squadrons continued to be kept busy, though ground targets were becoming the more usual objectives. The Mosquitoes of No. 23 Squadron, operating from Alghero, did make the occasional air-to-air kill, such as the destruction of an He 177 in the Istres area on the evening of 29 January by Flying Officer Grimwood in HX867/T. However, there was also a great deal of frustration:

> The night's programme was remarkable for some particularly misguided efforts on the part of the Sector Control at Alghero and Ajaccio, both of whom made blunders which left the aerodromes of southern France uncovered at times when hostile aircraft were almost certainly landing . . . Flt Sgt Shattock, when on his way to France, was picked up by Ajaccio Control and vectored all over the ocean in futile attempts to locate Huns.

January saw the squadron fly a record number of sorties, but with little to show for it. The following month the intruders were given a wider operational area that included the Gulf of Venice, and the first part of February was certainly livelier. Flight Sergeant Davidson was intercepted by a German nightfighter and a 30min running combat ensued in which neither aircraft could gain the upper hand. Then, all of a sudden, the German broke off the engagement. These operations were enhanced by the arrival of a detachment of No. 256 Squadron with their AI Mk. VIII-equipped aircraft. So it continued into spring, and the implementation of a new operation, Dolphin, with the aim of intercepting the reconnaissance aircraft that were operating at low level along the Spanish coast. As most of these sorties were flown by day, they are not relevant to this history.

After the near disaster of the Battle of Berlin, which to all intents ended with the raid of 30/31 January (although a few more such attacks were mounted up to the end of March), Bomber Command needed a short respite in which to recover and rethink its tactics. German nightfighter tactics during the Battle of Berlin had contributed in large measure to the RAF's loss of 384 aircraft. Employing spotter aircraft, usually Ju 88s, to track the bomber stream and call up other nightfighters had proved a success. This, combined with the use of fighter flares (Ju 88s dropping strings of flares around the bomber stream to indicate its position), had helped to reduce the effectiveness of the RAF's RCM operations.

Stuttgart was the target for the night of 15/16 March, 863 aircraft taking part. Other, smaller Bomber Command raids were sent against Amiens and Woippy, and this helped to split the German defences. Likewise, the tactic whereby the Main Force bomber stream flew south, almost to the Swiss border, before turning in towards the target caused some confusion among the nightfighter controllers and led to late intercepts. Most nightfighter activity thus took place over and around Stuttgart:

> Halifax 'X' of 76 Squadron was attacked just at the start of its bombing run over Stuttgart, at 20,000ft. Although the sky was dark above the bomber, cloud below reflected fires in the target area and made downward visibility fairly good. The flight engineer warned the captain that there was a single-engined aircraft on the starboard beam flying a parallel course at about 800 yards range. The engineer told the gunners that he would watch this aircraft while they continued their search. The captain meanwhile turned the Halifax slightly to port to make the bombing run, and immediately afterwards the engineer saw the fighter bank and, reporting this to the mid-gunner, told him to take over. The latter warned the captain to prepare to dive to starboard and gave him the word to do so when range had closed to about 500 yards. The fighter did not open fire, the speed of its approach and the bomber's manoeuvre doubtless making it impossible to do so. The mid-gunner, also, had no time to fire, but was able to recognise his opponent as an Me 109. In all, four unsuccessful attempts were made by the fighter to get into position to fire at the Halifax. Each time it came in the gunners and pilot co-operated skilfully to make it impossible for the enemy to carry out an attack.

The Messerschmitt came in for the fifth time on a 'curve of pursuit' attack, on starboard quarter rather above the Halifax. At 300 yards' range the tail gunner opened fire with two fairly long bursts. Strikes were observed on the fighter's fuselage. The fighter also fired as it started to close in, but scored no hits. The mid-gunner maintained his fire, and the enemy suddenly went down in a steep dive with smoke streaming from its engine. Its smoking descent was watched by captain, engineer and air-bomber, until it was lost to view as the Halifax turned to resume its bombing run.

During this engagement the Messerschmitt had come within 20 or 30 yards of the Halifax and while the mid-gunner was firing at it he was able to see plainly the black crosses on the underside of the wings, and the rest of the aircraft was coloured egg-shell blue. Thus it seems likely that 'X' of 76 Squadron had encountered and destroyed a day fighter which had been ordered to take part in the night defences of the Reich.

It was also in March that the Fleet Air Arm at last took its Fulmar nightfighters to sea, No. 784 Squadron attaching Flights to other units. B1 Flight to No. 813 Squadron on HMS *Campania*, B2 Flight to No. 825 Squadron on HMS *Vindex*, and B3 Flight to No. 835 Squadron on HMS *Nairana*.

Bomber Command was finding the business of night bombing equally dangerous. The Berlin campaign had been a shock, but worse was to follow. The Nuremberg raid of 30/31 March was nothing short of a disaster for the attackers. That night, 795 bombers (including 572 Lancasters) took off for the primary target, while other, much smaller formations prepared to attack decoy targets or undertake intruder missions. As this large raid began to build over England, the German reporting system was already passing the first of its situation reports, and the nightfighter units were brought to maximum alert. The bomber stream formed up and moved across the Channel, and the first of the nightfighters were waiting as they approached the Ruhr. Les Bartlett was bomb aimer in a No. 50 Squadron Lancaster:

> As we drew level with the south of the Ruhr valley, things began to happen. Enemy night fighters were all around us and, in no time at all, combats were taking place and aircraft were going down in flames on all sides. So serious was the situation that I remember looking at the other poor blighters and thinking to myself that it must be our turn next, just a question of time. A Lancaster appeared on our port beam converging, so we dropped 100ft or so to let him cross. He was only about 200 yards or so on our starboard beam when a string of cannon shells hit him and down he went.

That night the nightfighter force flew 246 sorties and claimed 101 enemy bombers destroyed, plus a further six probables, for the loss of five of its own aircraft. Royal Air Force records show losses of 95 aircraft, the majority of which were thought to have fallen to fighters rather than flak. The German controllers had not been fooled by any of the decoy techniques, and had ordered aircraft to orbit beacons Ida (Aachen) and Otto (Frankfurt), ready to be fed towards the stream. It was a clear night, almost perfect conditions for the defenders. The bombers made no effort to hide their approach and flew, unknowingly, straight towards the waiting fighters. The German controllers continued to give a commentary on the stream's course and height, but as soon as the first combats took place this was almost superfluous, as the night sky was alight with burning aircraft and the ground track was marked by burning wrecks.

Taking off from Mainz-Finthen at 2345, Hauptmann Martin Becker of 1/NJG6 made contact with the stream at about 0010. Ten minutes later he began a devastating attack on the Halifax bombers he was tracking, shooting down six within the space of 30min. He returned to base determined to rejoin the fight, and on his second sortie of the night caught up with the tail end of the homebound bombers near Luxembourg, shooting down his seventh victim. He was not the only one to make multiple claims. Hauptmann Martin Drewes claimed three and Leutnant Schulte four. All had used 'schräge Musik' to deadly effect.

According to the Bomber Command ORS of 3 May 1944:

> A substantial proportion of the bomber force is now equipped with either Monica III or Fishpond. A survey of the operational record has shown that both had some success in reducing losses and the number of attacks by night fighters. In the period Oct 43 to Feb 44 this is estimated as a 15 per cent reduction.

Many Bomber Command crews would not have agreed, as they considered the equipment a liability. Later in the year it would be withdrawn from operational use (except by 100 Group), proof having been obtained that the German nightfighters were equipped to home in on its signals.

Although the success rate for the Serrate-equipped aircraft had been steadily rising, and the intruder operations were paying dividends – albeit with a fairly high loss rate, the bombers were still too vulnerable. Nuremberg was nothing less than a crisis. Many more raids like that and Bomber Command would have to quit. It seemed that little could be done to improve the on-board defences of the bombers, other than to introduce the automatic gunlaying turret (AGLT) once this had been proved to be effective, so the only other option was to increase the scale of support operations. This policy was put into effect as rapidly as possible, with a growth in the RCM units and an increase in the number of Mosquito units. The importance of the electronic war cannot be too highly stressed, and the work of 100 Group deserves far greater attention than it has been given in recent years (although many of their records are still under security restrictions). With the apparent wisdom of hindsight some historians have suggested that Bomber Command should have reverted to daylight operations to take advantage of the increasing numbers of long-range day fighters available for escort work.

If the winter of 1943/44 had been a disaster for Bomber Command, it had also been a nightmare for the Luftwaffe nightfighter force. Increasing losses and a decrease in expertise were having a serious effect.

Among the Bomber Command raids on the night of 27/28 April 1944 was one by 322 Lancasters to Friedrichshafen, the target being chosen because parts for tank engines were produced there. It was a long haul across enemy territory, and although the target was on the edge of the nightfighter zones, the bombers were caught in the area of the target. Wilhelm Johnen and crew – including Paul Mahle, of 'schräge Musik' fame – were airborne in Bf 110 C9+ES, and before long had claimed one victim. As they moved in on a second Lancaster they were caught by return fire and lost the port engine. One of their victims was No. 35 Squadron Lancaster ND759, which managed to ditch in Lake Constance, the crew being interned by the Swiss authorities. However, the nightfighter was also in trouble, over Switzerland with alert Swiss defences and only one engine working. There was no option but to land at Dübendorf and face interment, not a prospect that Johnen or his crew relished. However, the Germans were desperate to get the aircraft back. The Swiss authorities saw the opportunity of making a good deal (as they had been doing throughout the war) and agreed to exchange the nightfighter for six new Bf 109G-6 fighters. The exchange was duly made, and the Bf 110 and its crew returned to Germany.

Most night actions had been taking place over occupied Europe, and bomber crews considered themselves to be safe once they had crossed the English coast. Then, as spring approached, the German intruders made a brief and devastating reappearance.

The heavy bombers of the US 8th Air Force were used to carrying out their missions in daylight, but on 22 April the timings were such that they returned to their bases as night was falling. It was a terrible mistake; the intruders of II/KG51 had followed them home. This unit had formed from a nucleus of V/KG2 in March, and at the end of that month had moved to

Soesterburg to carry out intruder operations with their Me 410s. This night in March was to witness their greatest success.

In mid-May the German nightfighter organisation was dealt a blow by Goering, who, having become concerned at the wastage rates among day fighter pilots, stated:

> To exercise economy in unit commanders, three Gruppen of single-engined night fighters are to be dissolved and both aircraft and pilots transferred to the remaining six single-engined night fighter Gruppen. At the same time, two Geschwader of single-engined night fighters are to be dissolved.

Although this had no effect upon the more important twin-engined nightfighter units, who were by far the most deadly as far as Bomber Command was concerned, it did reduce the overall effectiveness of the system by reducing the defences in the target area. What was to have a greater effect from this point onwards was the increased employment of nightfighters in the day fighter role, in an unsuccessful attempt to inflict crippling losses on the American bomber formations.

The RAF's intruder campaign continued through the spring and early summer, but was halted in late June so that the units could join in the battle over Normandy. It had been growing in effectiveness, as evidenced by this comment in the War Diary of 1 Jagdkorps for 19/20 February:

> As a prelude, about 100 Mosquitoes made a series of sorties into Holland, dropping bombs on the night fighter airfields of Deelen, Venlo, Leeuwarden and Gilze-Rijen; appreciable damage was done in places and some were out of action for 24 hours.

That night, Bomber Command had sent 823 bombers to attack Leipzig, with intruder support from only 16 Mosquitoes (plus a further 12 on Serrate patrols). The Luftwaffe flew 294 defensive sorties and claimed 82 contacts, leading to 74 kills, for the loss of 17 aircraft. Royal Air Force records show a total loss of 79 aircraft from all missions flown that night. This was the most costly night yet for Bomber Command, and included a loss rate of 9.5 per cent on the primary mission.

In March an American detachment, under Major Gates, joined No. 515 Squadron at Little Snoring in Norfolk for intruder operations, to test the suitability of the North American P-51 Mustang and Lockheed P-38 Lightning for such missions. Gates flew his P-38 on the first sortie, to the Berlin area, on 24/25 March. This and subsequent sorties revealed few problems, but brought equally few successes. Having flown some 60 operations by the end of April, the detachment left, no firm conclusion having been reached. Luftwaffe statistics for the period December 1943 to

April 1944 record 7,000 operational nightfighter sorties, claiming 1,100 victims – but this was still less than ten per cent of the bomber force. Furthermore, losses continued to be high, with an increasing danger of low-level day attacks on airfields.

In late April the Luftwaffe night raids switched from London to ports and shipping, in a belated attempt to frustrate Allied invasion plans. The raids remained small scale and ineffective, perhaps the only significant event being the operational introduction of the Fx 1400 'Fritz X' radio-controlled bomb, which was used against shipping at Plymouth. The use of such 'stand-off' weapons on a large scale would have obliged the defences to find and destroy the launch aircraft, as the weapons themselves were very difficult targets. In the event, however, the Germans were never in a position to undertake operations on such a scale.

The night defences maintained their earlier success rate against the normal night raiders, and by May all but two of the squadrons had re-equipped with the Mosquito.

The air plan for Operation Overlord, the Allied invasion of France, was one of the most complex military plans ever devised. One very small but important part dealt with the provision and operational deployment of the nightfighter force. It was considered that the Luftwaffe would be unable to attack the vulnerable landing areas by day, but that significant damage could be inflicted upon shipping and build-up areas by night. Among the many hundreds of ships involved with the landings were three Fighter Direction Tenders (FDTs), two of which, FDT 216 for the western area and FDT 217 for the eastern area, also had GCI controllers for nightfighter co-ordination. It was intended that this role would be taken over by the end of Day One by shore-based GCI units included as part of the early wave of landings.

The air assault began on the night of 5 June, as the invasion fleet moved into position. Nightfighters flew defensive patrols ahead and on the flanks of the fleet, No. 409 Squadron claiming a Ju 188. However, the task the following night was even more important, as the Germans now knew the landing points and would try to attack the mass of shipping, using torpedo aircraft and stand-off weapons such as the Henschel Hs 293 and Fx 1400. Once again the nightfighter squadrons took station around the area. Only one of the planned shore-based GCI sites was operational, in the British sector at Arromanches, and this undertook limited control of the 85 Group patrols. British estimates suggest a Luftwaffe effort of 175 aircraft that night, although only 40 approached the beach-head area, and of those the fighters claimed 12. Allied naval records note that three ships were damaged, including one LST that subsequently sank. Among the squadrons claiming success, No. 456 put in claims for four He 177s shot down in the area between Le Havre and Cherbourg, while No. 604 claimed five aircraft over France.

The six nightfighter squadrons of 85 Group were all involved in non-stop patrols over the area, combats taking place on most nights. By the end of June they had claimed some 60 enemy aircraft.

Thus the air situation over the beach-heads was secure day and night, and all seemed to be going according to plan. Then, in the early hours of June 14, the first of Hitler's 'retaliation weapons', the V1 flying bombs, landed on London. The next night the Germans launched 244 V1s, 73 of which reached London. A No. 605 Squadron crew shot down one of the weapons in the early hours of 15 June, and within days the day and night air defences had been reorganised to deal with this new threat. Among the units ordered into this new battle were the Mosquito-equipped Nos. 96, 219, 409 and 418 Squadrons. It was not an easy task. The targets were small and fast, and judgement of range was difficult. By mid-July there were some nine Mosquito squadrons on such 'anti-Diver' patrols.

By day, the Hawker Tempest was proving itself adept at V1 hunting, so the 'powers that be' took the logical step and proposed use of the aircraft at night. In August No. 501 Squadron moved to Manston in Kent and re-formed as a nightfighter unit to operate against V1 night attacks, the brief being to cover the Kent coast to the North Downs. To provide the unit with appropriate experience, pilots were drafted in from the FIU and selected nightfighter units, although a number of experienced Tempest men remained on strength. The new unit was put in the hands of Squadron Leader Joe Berry, a V1 ace experienced in night operations against these weapons (his score included seven shot down on the night of 23/24 July).

One of the pilots at this time was Flight Lieutenant Jimmie Grottick, who recalls:

> The pattern soon became reasonably settled, that is so far as proper patrol lines for the V1 were concerned, especially at night. Mainly we came in at line astern with the hope that we could judge 200-300 yards and open fire. Later they installed a bit of radar into the aircraft so that at the point of overtaking the target a little orange light came on in the cockpit to warn that you were both 'in range' and 'getting too close'.

This device had been developed at short notice by the TRE, under the guidance of Sir Thomas Merton. In essence, it was a simple spectroscope with a refracting system that caused the exhaust glow of the V1 to appear as two images until, at a range of 200 yards, they merged into a single image. The V1s were never easy targets, day or night, and many attacking aircraft were damaged or destroyed when a V1 exploded upon been hit.

Then, in early July, the Germans added a new dimension to the campaign by using He 111s as launch platforms for V1s, this move being in large part due to the loss of land-based sites as Allied ground forces advanced. Because the launch aircraft flew at slow speed and at low level on dark nights, they were not easy to find. A radar picket ship, HMS *Caicos*, and a radar-equipped Wellington were used to try and improve detection rates. These He 111s became a priority target for the nightfighters, but the first confirmed success did not occur until 25 September (by No. 409 Squadron). Dick Leggett flew a Mosquito on such missions with No. 125 Squadron:

> Once the ram-jet of the flying bomb had ignited, the pilot [of the Heinkel] would quickly descend to his original height and head for home, using low stratus or sea fog to cover his retreat. During the few minutes taken to launch the missile the enemy aircraft was vulnerable to attack from the numerous nightfighters which would always be in the area due to the prior warning given to us from intelligence sources. However, as winter set in, the German pilots became more adept at cutting their exposure time to a minimum, thereby avoiding interception during the launching period.
>
> It was under these circumstances that Operation Vapour was evolved, in early 1945, so that the Heinkels could be intercepted and destroyed whilst at low altitude. Prior to this tactic, life had been more difficult as the Heinkel could happily cruise at a speed well below the stalling speed of a Mosquito XVII. On one particular intercept, each time visual contact was attempted the severe turbulence from the slipstream of the Heinkel destabilised my aircraft and created an incipient stall.
>
> Throughout the interception my nav/rad, Egbert Midlane, miraculously held radar contact with the target and repeatedly guided me into his minimum AI range of 100 yards, but cloud and darkness prevented a visual sighting. We decided to wait for the greyness of dawn and were eventually rewarded by the destruction of the Heinkel as it approached Den Helder. Some 55min had elapsed from first radar contact to firing the four 20mm cannon.

The campaign against these aircraft was joined by a number of other units, including a detachment of Firefly NF.IIs from No. 746 Squadron. In general, the defenders had responded rapidly and very efficiently to the V1 threat, and although many of these weapons continued to fall on England, the majority of them were claimed en route by aircraft and guns.

FAIREY FIREFLY

The prototype Firefly first flew in December 1941, its primary role being that of day fighter-reconnaissance, and it duly entered service with No. 1770 Squadron in October 1943. However, by mid-1941 the Admiralty had recognised the need for a specialist nightfighter and, after successful trials with a modified Fairey Fulmar, started talks with Fairey for the production of a nightfighter variant of the Firefly. The first such NF.I aircraft were produced from the basic F.I by adding an ASH radar (basically an American AN/APS 4) in a nacelle under the forward fuselage. The NF.II was produced along similar lines, although very few of the 328 ordered were completed. This was followed by the NF.IV, which carried the ASH in a nacelle on the port wing, with a fuel pod under the starboard wing for balance. The final variant, and the one that became the definitive Firefly nightfighter, was the NF.5, which saw active service in the Korean War with No. 810 Squadron.

Firefly NF.5 data
Powerplant: One Rolls-Royce Griffon 74. Crew: Two. Length: 37ft 11in. Wingspan: 41ft 2in. Maximum speed: 386mph. Ceiling: 29,200ft. Armament: 4 x 20mm cannon.

Having gone somewhat ahead, chronologically, the story must take a pace backwards. A German Staff Memo dated 2 July 1944 stated:

> The Führer, wishing to protect the homeland and secure the front, issued the order that fighters are to have absolute top priority and heavy combat aircraft are to be cancelled. The stock of fighters, destroyers and night fighters is to be increased to 10,000. This number is to include 5,000 new machines – including 500 night fighters.

Despite the Allied bomber effort against German aircraft production, the factories continued to turn out large numbers of aircraft. In the period from March to December the Luftwaffe received 3,492 nightfighters (out of a production total of 3,886 in the same period). The same Staff Memo also cancelled one of the new breed of German nightfighters, the Focke-Wulf Ta 154. As mentioned previously, the first production version of Kurt Tank's primarily wooden aircraft had flown in June, but was still suffering severe problems. It was decided to cancel the type in favour of the successful Do 217.

The German nightfighter force suffered another blow on 13 July when Ju 88G-1 4R+UR of 7/NJG2, equipped with all the latest devices, landed at Woodbridge in Suffolk after the crew (Obergefreiter Mäckle) flew a reciprocal compass heading and become hopelessly lost. Once again, as with the aircraft that had landed at Dyce in May 1943, the RAF was able to acquire an intimate knowledge of German capabilities (including SN-2, Naxos and Flensburg), and so was able to develop a new range of countermeasures. The German crew had taken off from Volkel in Holland for a routine night patrol, and at some stage had made the navigational error that caused them to head for England. Unable to pinpoint their position, and concerned about their low fuel state, they landed at the first set of airfield lights they found, assuming that they were somewhere near the Dutch-German border. Only when they climbed out of the aircraft, having parked next to a Stirling, did they realise their mistake.

The aircraft soon acquired RAF markings and the serial TP190, and was flown to the Royal Aircraft Establishment at Farnborough for extensive testing. Meanwhile, the scientists were presented with the latest FuG 220 Lichtenstein SN2 and FuG 227 Flensburg to study. The latter at last proved that the Germans had the ability to home on to the Monica transmissions, a discovery that led to the virtual abandonment of this equipment.

By mid-summer the nightfighter force comprised some 15 per cent of total Luftwaffe front-line strength. Overall fighter strength was roughly equally split between day and nightfighter elements, although the former did not suffer the same serviceability problems. The night skies were an arena for a deadly game of cat and mouse. If the nightfighters managed to find the bomber stream, carnage would result and burning bombers would litter the countryside. If, on the other hand, the RAF could send the defenders to the wrong target or cause major disruption to his airfields and control facilities, then losses could be kept down. By early 1944 the German defenders had seen most of Bomber Command's tactics. Experienced controllers were able to work through the electronic countermeasures and marshal their resources until the real target for Main Force had become clear.

Such was the case on the night of 18/19 July, when Bomber Command flew 927 sorties to a variety of targets. On most raids the losses were low, but this was not the case for the attacks against the marshalling yards at Aulnoye and Revigny. The total force employed against these two targets was 253 Lancasters and 10 Mosquitoes, the intention being to destroy rail communications to the Normandy area as part of the strategic interdiction campaign supporting the invasion. The Aulnoye raid went reasonably well, only two Lancasters being lost. However, the 5 Group attack against Revigny was caught by nightfighters and lost 24 aircraft, almost 22 per cent of the Lancaster force involved. No. 619 Squadron from Dunholme Lodge, Lincolnshire, had a particularly bad night, losing five of their thirteen aircraft.

Among the nightfighter pilots airborne that night was Herbert Altner of III/NJG5, who used his 'schräge Musik' to shoot down five bombers in just over 30min. Altner was an experienced nightfighter pilot, having first joined NJG3 in 1943, and had perfected an attack profile that would destroy the aircraft but, he hoped, enable the crew to escape. After shooting down the first Lancaster by aiming between the port engines, he '. . . broke off to the left down and levelled out. Because the approximate course of the formation was known to me, I pulled up again, and the other attacks were straight out of the textbook. After the fifth I'd had enough and flew home.' While 'swimming with the stream' made multiple scores possible, it also put the nightfighter at risk of colliding with a bomber or being hit by an unseen yet alert air gunner.

This period, however, was to be the peak of the nightfighter force's achievements, and from this point on it would be in decline.

Meanwhile, July saw the European operational debut of the long-awaited Northrop P-61 Black Widow. In the late 1940s Lieutenant-General Delos Emmons had pressed for a nightfighter design to be put into production, to include a 'device for locating enemy aircraft in the dark'. The Northrop design, designated XP-61, was chosen and an initial contract was issued in March 1941. The aircraft entered service with the 458th Night Fighter OTU at Orlando Field, Florida, in July 1943, the 348th NFS becoming operational as an air defence unit for that state.

There was a desperate need for reinforcement of the nightfighter force on the European Continent. The six squadrons of 85 Group had been working hard, scoring successes and also taking losses, but the continued V1 offensive and a fear of major bomber attacks meant that no other UK-based units would be transferred in the near future. The first P-61 unit, the 422nd NFS, arrived in the UK in February 1944 to join the 9th Air Force. However, owing to various delays in the arrival of their aircraft, the crews were detached to RAF squadrons to 'learn the ropes', enduring a great deal of frustration at being unable to get on with the job. The first aircraft arrived at Scorton, North Yorkshire, in May, and so began an intensive period of training, often in co-operation with local Halifax units. By June, detachments of the 422nd and the newly-arrived 425th were at Hurn, ready for action. The first mission, a patrol along the Normandy coast, was flown on 3 July by Lieutenant Colonel Johnson in P-61A *No Love, No Nothing.* The following month the 422nd was on detachment at Ford as part of the air defence against V1s, and it was here that they chalked up their first 'kill' (the first victory against a piloted aircraft was not scored until 7 August, when a Ju 88 was shot down near Mont St Michel). In late July the unit began a move to a new base in France, becoming operational in early August. This gave the AEAF three nightfighter squadrons in France; the American 422nd and Nos. 264 and 604 Squadrons RAF, both equipped with Mosquito XII/XIIIs.

Other than limited experience in North Africa and during the D-Day period, the RAF had not as yet been called upon to provide nightfighter cover of the battle area. Now, as the Allied troops advanced across Europe, the Luftwaffe began to undertake night bombing of the ground forces. The 2nd Tactical Air Force (TAF) saw problems in the forward employment of the nightfighter units:

NORTHROP P-61 BLACK WIDOW

The Northrop P-61, the first purpose-built American nightfighter, was designed in response to a US Army Air Force requirement of late 1940 for a twin-engined aircraft with two crew and 'a device for locating aircraft in the dark'. The initial proposal was put forward by Northrop in November (at the same time that Douglas put forward its XA-26A), and in mid-December the go-ahead was given for a prototype of the Northrop NS-8A, as the XP-61. Development work was still under way when the first production contract was issued in March 1941; an indication of the urgent need for the aircraft. Work continued on the aircraft and systems over the next twelve months, the SCR.720 being seen as the standard AI radar, and the first flight took place on 26 May 1942. By early 1942 orders had been placed for over 1,700 P-61 'Black Widows'.

The production P-61A included a four-gun dorsal turret in addition to the major firepower of four 20mm cannon, although this was subsequently removed owing to a shortage of turrets and a problem with tail buffet. The first deliveries were to the 481st Night Fighter Operational Training Unit at Orlando Field and the 348th Night Fighter Squadron. Other units soon followed, and as soon as basic training was complete they were shipped to overseas operational theatres in Europe and the Pacific. One P-61, 42-5496, was sent to the UK in March 1944 for assessment by the RAF and trials with the Mosquito nightfighters.

The P-61B was given the improved SCR.720B with longer detection range, and was also fitted with night binoculars linked to the optical gunsight. The final variant was the P-61C, powered by supercharged Pratt & Whitney R-2800Cs in an attempt to improve altitude and speed performance. Production orders were issued for more than 400 of this variant, but only 54 had been built before the order was cancelled at the end of the war.

The two operational variants saw extensive service and achieved a number of air-to-air kills. They were also employed on night ground-attack duties, being fitted with bomb racks to carry a variety of weapons.

P-61A data

Powerplant: Two Pratt & Whitney R-2800-10/65. Crew: Three. Length: 48ft 10in. Wingspan: 66ft. Maximum speed: 372mph. Ceiling: 34,000ft. Armament: 4 x 20mm cannon and (with turret) 4 x .50 machine-guns.

> Nightfighter squadrons require airfield and maintenance facilities which are seldom available in forward areas during mobile operations. It is best to retain night fighter squadrons under command of the AOC of the Base Group, but it should be possible to detach squadrons to forward airfields and place them under control of Tactical Group commanders should the need arise. Although the effectiveness of air attack by night in forward areas should not be underestimated, it is, nevertheless, a fact that a night bomber force of any size is a strategic weapon more likely to be used against ports and base installations than against units in the front line.

Meanwhile, operations in the Mediterranean and Italian theatres had increased in scale as Allied armies moved further up Italy and Allied bombers used Italian bases to launch attacks against southern Germany. In July the majority of nightfighter squadrons, including the American 414th to 417th NFSs, still operated Beaufighter VIs, although two Mosquito units, Nos. 256 and 108 Squadrons, were also active. Although air combats did occasionally take place, the nightfighter units now tended to spend more time employed on night ground attack.

On the night of 29/30 August, Bomber Command split its effort between Stettin (402 Lancasters) and Königsberg (189 Lancasters). Among the aircraft on the Stettin raid was ME742/B of No. 626 Squadron. Flying Officer Hawkes recalls the event, his 25th mission:

> 0021 fix, on track, on time. 14 miles from Swedish coast, 13,000ft, 098, climbing to 18,000ft. We were attacked without warning by two Ju 88s. Fishpond was fitted but was not being monitored. Both fighters opened fire from dead astern with cannon tracer, no crewmen were seriously injured but the hydraulics to both turrets severed. Only the mid-upper could be rotated manually. There was a large hole in the starboard wing and fire in both starboard engines. The elevators and rudders were holed but we still had enough control to corkscrew. The Flight Engineer, Sergeant Ockwell, threw window. I feathered the starboard outer but the inner would not feather. Ten seconds later fire broke out again and spread to the No. 1 petrol tank. The mid-upper gunner, Flight Sergeant Allison, saw a Ju 88 on a parallel course so opened fire – it burst into flames and went down through the cloud. I levelled the aircraft at 9,000ft and headed for Sweden. The fire had well and truly spread and there was a danger of the main spar giving way so I ordered the crew to abandon aircraft.

All of the crew escaped from the doomed Lancaster.

An account of this combat was included in the *Bomber Command Quarterly Review* for July–September 1944:

> The moonlit night was clear above the patchy cloud which spread across the sky at 19,000ft, and the enemy flares which lighted the way of our bombers

were easily seen. The target that night was Stettin; the route to it lay over that narrow strip of water dividing Sweden and Denmark at their closest approach, the Sound, which Lancaster 'B2' of 626 Squadron was crossing some 3,000ft above the cloud. Danger of interception by fighters was advertised by the flares, and a good look-out was kept to make sure that no attacker could sneak into lethal range unobserved.

It is the business of the night fighter to make his attack from as short a range as possible, taking his opponent unawares. From the moment when he sights his quarry, therefore, he must try to remain unobserved. In this, naturally, he is not always successful; although occasionally he succeeds all too well. For all the care of her gunners, 'B2' was attacked and hit by a Ju 88 which, unseen, had approached from directly below the starboard quarter to within very close range. Almost immediately another Ju 88 was reported by the mid-upper gunner to be closing in from the same direction, although without opening fire.

The first Ju 88 raked the bomber from below and set her on fire, smashed the hand crank of the rear turret, wrecked the mid-upper gunner's sighting bar and blasted off the Perspex. In this condition the aircraft's plight was sorry enough, but a second attack developed from the starboard quarter. An explosion in the starboard wing; a fire in No. 2 tank; the cutting-out of both starboard engines, these were the immediate effects. Petrol was soon pouring from the damaged tank over the whole starboard wing, which was ablaze; but by good fortune the incendiary load in the bomb bay did not ignite. Despite damage to the rudders and elevators, the captain managed to carry out a modified 'corkscrew' in response to the gunner's commentary. The aircraft, flying at 9,000ft, was now within two minutes of the Swedish coast and the captain warned the crew that they must bale out, the fire was now spreading to the fuselage.

Although one fighter had broken away and disappeared from view, the other flew alongside the Lancaster at about 50 to 100 yards' range on the port beam. Handicapped by the damage to his gun, which deprived him of his gun sight, the mid-upper gunner succeeded in firing a burst of some 400 rounds. So accurate was this fire that the Ju 88, bursting into flames, rolled over and disintegrated as it fell into cloud. The Lancaster having thus been avenged, the gunner followed the wireless operator and rear gunner, who had already baled out through the rear door.

As Russian forces advanced in the east, they made increasing use of air power to attack German cities, including Berlin, and troop concentrations. The majority of bombing was carried out by night, as the Luftwaffe still dominated in the daylight hours; Soviet ground-attack aircraft suffered heavy losses to fighters and flak. Although the German night defences had been strengthened, they did not receive the same priority as those in the western regions. In an effort to strengthen them further, a limited number of Bf 110F and G variants were transferred to other Axis air forces in the eastern region, the 101st Night Fighter Flight of the Hungarian Air Force, operating out of Csorne, being an example.

By the latter months of 1944 the Allied nightfighter forces operating over Europe were confined to flying standing patrols along sectors of the front line, under GCI control, and were finding all too few 'customers'. When an intercept did take place, the 'target' was invariably an Allied aircraft in the wrong place.

However, the night hours were still being used by the German army to move men and material around the front line, daytime being too well covered by Allied fighter-bombers to permit safe movement. The scale of 'nightfighter' operations against ground targets, and of Mosquito squadrons' operations against rail targets in particular, increased throughout the year. The P-61s joined the intruder game, using rockets and 260lb fragmentation bombs. The ground radar units vectored the intruders towards the area, and flare-equipped aircraft then illuminated the target.

A shortage of AI Mk. X led to the use of the Mk. XV, derived from the American 3cm Army-Navy Airborne Pulsed Search (AN/APS) 4 set, in 100 Group Mosquitoes, as an interim measure.

In a comprehensive document dated 5 November, 8th Abteilung summarised the 'problems of German Air Defence'. Sections of the document addressed the nightfighter question:

> As regards night fighters, the position is rather more favourable at present. Substantial reinforcements have been received from disbanded bomber and transport units, and striking power has, therefore, greatly increased. Our total strength of approximately 1,800 aircraft enabled about 200 fighters to take the air during each enemy attack. Night fighter crews have been trained on an almost peace-time scale, and successes have been considerable.
>
> On the night of November 22/23 the first raid under bad weather conditions was effected. Our night fighter operations were severely prejudiced by the weather. The success of Himmelbett was very largely dependent on the strength of enemy jamming, but reasonably good results were achieved by the use of 'Wilde Sau'.
>
> A plan must be devised for the concentration of our forces in western Germany. In this connection, the following suggestion may be borne in mind. Since the present fuel shortage only permits of the employment of night fighters for a few days each month, our forces should carry out operations at full strength on certain days and times as based on previous experience. Although a certain percentage of such sorties would be fruitless, it can be assumed that 50 per cent will succeed in contacting the enemy. This is the only way in which immediate protection can be given to the heavily bombed areas of western Germany.

A further solution would be to convert some of our night fighter units to Me 262s. This would enable us to attack and inflict heavy losses on the Mosquito squadrons which are operating in ever increasing strength.

One of the major problems had been the reduction in effectiveness of the Air Reporting Service, in large part due to the loss of territory in northern Europe – in other words, the loss of forward radar coverage that had previously provided accurate information enabling the nightfighters to begin their attacks even before the bombers had crossed the coast. While the numbers of fighters continued to increase, all other aspects were in decline. Fuel shortages limited crew training and even forced operational restrictions on the front-line units, and on many nights Bomber Command went virtually unhindered because certain nightfighter units were not permitted to fly. Furthermore, as far as the situation in the west was concerned, the transfer of units to the Eastern Front to counter Russian bombing attacks further depleted operational strength. In view of these and other factors, it is remarkable that the nightfighter force remained as effective as it did.

On 10 November a 'summary of the war situation' was published, part of which concerned nightfighter operations:

The following are essential for night fighter attacks:

1. Extension of night fighter control aids.
2. Intensification of radio warfare with the aim of restricting enemy W/T and radar over Germany.
3. Increased operational use of close night fighter control (Himmelbett system) as a precautionary measure in western areas.
4. Operational use of the Me 262 and Do 335 against Mosquito formations.
5. Airborne radar equipment showing on one screen one's own position and that of the enemy aircraft, thus giving greater freedom from ground control, which is liable to interference.
6. Installation of a homing device with continually changing frequencies to counteract enemy jamming.
7. Coupling of the radar scanner with the armament for blind firing.
8. Parking of night fighters in rain and snow-proof hangars.

A number of these elements were already being addressed; others would prove beyond reach before the end of the war. Technical improvements appeared almost monthly as new equipments were brought into service, including the FuG 218 Neptun. This system used variable frequencies in the 158-187MHz band, giving a maximum range of 3 miles and a minimum of 350ft.

The operational debut of the Messerschmitt Me 262 jet fighter came as a major shock to the Allies, but the only aspect of this aircraft of concern here is its limited employment as a nightfighter. In November, Major Gerhard Stamp was given command of a number of experienced 'Wilde Sau' pilots and, as 'Kommando Stamp', was provided with a small number of Me 262B-1a development aircraft (thought to be ten). One of the highest priorities for the unit was countering the Mosquito intruder operations.

By November, the RAF's premier intruder unit, No. 23 Squadron, had been back in the UK for six months, carrying out the same effective campaign against air and ground targets for which it had justly become famous. The squadron began sending crews on courses to learn the ins and outs of Air to Surface H (ASH, the American AN/APS 4), and by December were operational with this equipment: 'In the early noon, another "gen" man of the Royal Navy called on us to give the crews the griff on the ASH' – as the Navy used this radar, they were seen as the acknowledged 'experts'. The first ASH operation was flown on 22 December by Mosquito VI RS507; the crew obtained one ASH contact but it came to naught. The ASH aircraft were confined to 'A' Flight, who were thus 'affectionately' known as the 'ASH cans'. The first confirmed victory was scored on New Year's Day 1945 by Squadron Leader J. Tweesdale and Flight Lieutenant L. Cunningham in RS507, the victim being a Ju 88 found at Ahlhorn: 'After an 8 minute chase the Mosquito closed to visual contact, a 5sec burst was fired at 600ft, strikes seen all over the belly and port engine burst into flames. Another 5sec burst and the Ju 88 spiralled into the ground.'

The German offensive launched through the 'impenetrable' Ardennes on 16 December caught the American ground forces off guard. Bad weather prevented the use of the Allies' overwhelming air power, and for a while the situation looked critical. Even when the weather improved there was a shortage of nightfighters to counter the Luftwaffe resupply missions.

In the Pacific theatre, January 1944 saw the first carrier assignment of the US Navy's first nightfighter, when Vought F4U-2s (mod) of VF(N)-101 embarked on USS *Enterprise*. This first nightfighter variant of the F4U Corsair was originally proposed in November 1941, when the US Bureau of Aeronautics (BuAer) suggested that fifty Vought-built Corsairs should be adapted into nightfighters by the installation of a radar (in the starboard wing) and other night-flying instrumentation. Initial modification work was carried out at the Philadelphia Naval Aircraft Factory, additional work being done at NAS Quonset Point. The F4U-2 variant (of which 34 were eventually produced) was used by VMF(N)-532, VF(N)-75 and VF(N)-101, and entered service in mid-1943.

By the early months of 1944 the Americans were ready to launch the next stage of their re-conquest of the Pacific. On 31 January the Marshall Islands campaign began, launched by a task force of almost 300 ships, supported by 700 carrier-based aircraft and 475 land-based aircraft. The primary Japanese air base was Rabaul and its associated ancillary airfields. American bombers were flying an increasing number of night raids, and the Japanese maintained a limited capability in the same field; therefore the nightfighter forces of both sides were strengthened. At Rabaul in March 1944, Hikotai 316 became a nightfighter unit equipped with the Mitsubishi Zero, a type that had already proved its unsuitability for the role. Most of the pilots were ex-two-seat floatplane men who, after a quick conversion to the Zero at Atsugi, were expected to become day and night fighter pilots. Despite the notional nightfighter tag for the unit, it was committed to the day battle and was shot to pieces. Although a number of other units were also formed in the area, again with a night role, most met with little success and suffered catastrophic losses when thrown into the daylight battle.

Although Japanese night bomber operations were very restricted, the so-called 'Washing Machine Charlies' were a nuisance that the Americans wished to counter. Among the units deployed to carry out this task, VMF(N)-532 operated out of Tarawa in early 1944 and VF(N)-75, which also had the distinction of having been the Navy's first nightfighter unit, went to Bougainville. The 418th NFS was active as part of 5th Air Force, and although air combats were few and far between, there were exceptions. On the night of 29/30 December, Carroll Smith and his radar operator, Phillip Porter, flew two sorties, and claimed two enemy aircraft on each one.

The increased weight of air attacks against major installations in Japan and other critical areas led to an effort to improve the night defences. Among the specialist units was Air Group 302, formed in March to defend the Yokosuka base area and given a strength (on paper) of 48 day fighters and 24 nightfighters. The unit used a number of modified airframes with angled guns, including Zeros, Raidens and Gekkos. The primary problem remained that of mis-employment, with the nightfighters being used on other tasks.

The first Boeing B-29 raid against the Japanese mainland took place on 15/16 June and was aimed at the Yawata iron and steel works, the bombers flying from bases in China. This was one of the first operational nights for the home defence nightfighter units, such as the 4th Army Air Group (Sentai) which operated the Kawasaki Ki-45 Toryu (Dragon Killer). This twin-engined aircraft was a failed long-range escort fighter hastily converted to the nightfighter role and given a variety of weapon fits, including a 37mm anti-tank cannon for use as a B-29 destroyer. Among the defenders airborne that night was Isamu Kashiide. Having located the bombers, he moved in to the attack, claiming to have scored a direct hit with his cannon. The total Japanese claim for the night was five shot down and five damaged, but American records show no losses and no damage.

The year 1944 also saw the introduction of other types to supplement the Gekkos; the Yokosuka D4Y2-S Suisei (Comet), codenamed 'Judy', and the Yokosuka P1Y1-S Byakko (White Light) and P1Y2-S Kyokko (Aurora), both Ginga variants codenamed 'Frances'. By late 1944 the main enemy aircraft over the homeland was the B-29, but the defences remained ineffectual, with small numbers of aircraft and inexperienced crews. Units had impressive names, however. Air Group 352 was 'Kusanagi' (Heavenly Grass-Mowing Sword); this particular unit was formed in August 1944 for defence of the Nagasaki area and, in common with most such Air Groups, had day and night elements.

The Japanese certainly undertook a great many trials weapon fits. A US Strategic Bombing Survey entitled *Japanese Air Weapons and Tactics* gave details of a number of such installations on nightfighter types:

Ginga (Frances) – Four 20mm forward firing guns inclined upwards at 30 deg. Seventeen 20mm guns in the bomb bay, 12 firing forward, 5 rearward and all angled downwards. [This was for attacking ground targets such as airfields on night intruder missions, and was only an experimental fit.]

Raiden (Jack 32) – Two 20mm guns inclined at 30 deg fore-upwards and two 20mm wing guns.

Suisei (Judy 12) – One 20mm gun inclined fore-upward plus two 7.7mm wing guns.

Reisen (Zeke) – One 20mm gun inclined at 30 deg fore-upwards plus two 20mm wing guns.

Tenrai – Two 20mm guns, plus two 13mm guns, all inclined fore-upwards.

Tactics – Antibomber – The Gekko was used operationally with upward, downward and sideward guns, but battle experience proved that the upward firing installations were most effective. The Gekko was successfully employed against B-29s and heavy bombers. Future plans were centred on

> wider use of 30mm cannons [*sic*] in inclined mounts. Airdrome attacks – 30 Ginkas with 17 20mm guns in downward inclined installation were in preparation to attack B-29 bases in the Marianas. The Navy believed that one pass could wipe out all the B-29s arranged alongside the runway, but the untimely termination of the war put an end to this scheme.

The final months of the war were to see the two remaining Axis partners battling against the ever-increasing weight of Allied air power. It was a losing battle, but the defenders never abandoned their task.

Chapter 8

THE NIGHT CONFLICT REMAINS FIERCE

The effectiveness of the German Ardennes Offensive of December 1944 demonstrated that the war was far from over. Launched in the early hours of 16 December, the attack caught the Allies off-guard, making early gains in the attempted sweep towards Antwerp. The overwhelming weight of Allied air power could not be brought into play because weather conditions were appalling. As the offensive met increased resistance, so aircraft began to play a greater part. The only aspect of concern here is that highlighted by an American postwar report stating:

> The night fighter strength allocated to the European Theatre was decidedly inadequate to meet the minimum requirement, as was amply demonstrated during the Ardennes offensive, when the German Air Force operated between 30–80 supply aircraft nightly without serious interference.

For the P-61s of the 422nd NFS, this was to be their most active period, and aircraft were heavily tasked both with defensive patrols and air-to-ground missions. Despite weather restrictions, the unit operated at every opportunity, eliciting great praise from the land force commanders and being awarded a Distinguished Unit Citation (DUC). The same period also saw the creation of the first 'ace' crew for the unit, Lieutenants Paul Smith and Robert Tierney scoring their fifth victory in late December 1944.

Meanwhile, the P-61 also entered service with 12th Air Force in Italy, four nightfighter units (the 414th, 415th, 416th and 417th) receiving the type. However, lack of aerial targets, and the needs of the ground war, meant that missions were mainly confined to night ground attack.

The training of RAF crews continued at No. 54 OTU, although the throughput of students had declined during the latter part of 1944. The importance of the course remained high, and as techniques, equipment and facilities improved so, too, did the quality of the graduates. One major advance had been the employment of five Wellington XVIIIs fitted with AI Mk. X for use as flying classrooms. Each aircraft had five or six student consoles and one or two instructors, the students taking it in turns to direct the 'fighter' Wellington. Targets were usually provided by the unit's own

Hurricanes. The major advantage of this system was that an experienced operator could observe and instruct in a real scenario, rather than in a theoretical classroom situation.

With the German bomber force undertaking very few attacks on England, the major offensive remained that by the V1s. The He 111 launch aircraft were still causing problems. Dick Leggett was one of the No. 125 Squadron pilots involved in a special type of mission:

> We were pleased to learn that the boffins were planning a possible answer to the menacing Heinkels and their under-slung V1 missiles. On January 2, Midi [Egbert Midland, his navigator/radar operator] and I were sent to join Operation Vapour at Ford. Immediately on arrival we were ushered into a briefing room to meet the head boffin, Mr E. J. Smith. He then introduced us to the captain of the Coastal Command Wellington that was to act as an airborne GCI station – with Mr Smith as controller. After a thorough briefing we and five other night fighter aircraft joined up with the Wellington in our first daylight trial over the English Channel. The Wellington cruised at 150ft while we fighters were positioned long line-astern and stepped-up at intervals of a few hundred feet. Our navigators could position quite accurately by monitoring special signals from the Wellington. We were impressed by the accuracy of the interception vectors being given by Mr Smith, directing the fighters towards evading target aircraft at minimum altitude.
>
> Therefore, it was with great confidence on January 14 that the 'Vapour Circus' took up action stations at Manston to settle our fight with the Heinkels. As dusk descended, intelligence from 'Y' Service sources forecast that 40 or 50 Heinkels would arrive in the usual area off the Norfolk coast . . . Shortly after joining up with the Wellington we were told that Mr Smith had trade for us. Under his excellent close control we reduced height to 100ft, using our radio altimeter for complete accuracy, and followed his courses-to-steer at a speed of about 200mph. Within a few minutes, Midi obtained a firm contact on his AI and took over. Speed was synchronised with the target on a course towards Norfolk. With my gun button to 'fire' we struggled through the severe downwash of slipstream from the target and quickly established a visual sighting . . . closing to 100 yards . . . but the target turned out to be a Warwick. As the 'target' was not responding to IFF, Mr Smith ordered it shot down, but having definitely identified it as a Warwick, we left it alone. The Circus patrolled for the rest of the night but with no luck.

Despite the lack of success against Heinkels, the Warwick interception had proved the system. The Circus was kept in being for a few more weeks, but with equal lack of success. It was an advanced concept but, surprisingly, was not followed up.

With the loss of the launch pads in the Low Countries, the German V1 offensive relied on these airborne launches, but they could never be more than a nuisance; their numbers were simply too restricted. The last V1 fell on the UK on 29 March.

At long last the Royal Navy had formed its first operational nightfighter squadron, the various trials, detachments and so on having provided the basic concepts and techniques for employment. The Naval Night Fighter Interception Unit (NNFIU) had visited the US Navy development centre at Quonset Point, Rhode Island, in late 1943 to look at radar development of the AN/APS 4, or ASH, and were impressed by it compact size and overall performance. By mid-1944 ASH sets had been made available for fitting to Fairey Firefly trials aircraft. Six Fireflies formed the initial complement of No. 1790 Squadron, under the command of LieutenantiCommander J. H. Neale, on 1 January. Crews completed a two-month course at Drem, in East Lothian, with No. 784 Squadron, carrying out an intensive programme of night flying, navigation and practice interceptions. It was not until 24 May that twelve aircraft joined HMS *Puncher* for final work-up deck trials. As two aircraft crashed on landing, the deployment was postponed until the following day, when, in better conditions, the remaining aircraft landed safely on the carrier. It was considered that the crews were now qualified, and the squadron flew new aircraft to Sydenham, ready for embarkation on HMS *Vindex* and transfer to the Far East for the war against Japan. The unit eventually arrived in Australia in August, too late for the war; nevertheless, they remained part of the 8th Carrier Air Group.

In the meantime, there had been other RN developments, such as the acquisition of AN/APS 6 equipped Grumman Hellcats in January 1945. These went first to No. 892 Squadron under Major J. Armour, who had been a member of the team that had visited Quonset Point. A second unit also received Hellcats, but neither unit was operational before the end of the war.

In Europe, despite the increasing weight of Allied air attacks and the restrictions on supplies of vital aircraft parts and, more importantly, aviation fuel, the Luftwaffe nightfighters were still a force to be reckoned with. The 'massive blow' that Schmid had outlined on his appointment as General der Nachtjagd was simply awaiting the organisation of a sufficient number of nightfighters and the right opportunity. As early as November, units had been warned to expect 'Zeppelin' (the code warning for Gisella, the mass intruder mission) at any time. When the Ju 88 of Unteroffizier Lattoch landed in Luxembourg on 1 January, the Allies found copies of the plan for this operation, but were not completely convinced of its authen-

GRUMMAN F6F HELLCAT

The XF6F-1 was designed as a day fighter for carrier operations, and first flew in June 1942. Entry to service was rapid, as there was an urgent need for a carrier-based aircraft capable of matching the Japanese Zero. The F6F-3N was the first of the nightfighter variants, making its first flight in 1943. This variant had an APS 6 radar under the starboard wing and an armament of six .50 Browning machine-guns in the wings. While the US Navy was the major user of this type, a number were supplied to the Royal Navy as the Hellcat I NF. The F6F-3E was a sub-variant equipped with a modified radar. However, it was the next variant that was produced in the greatest numbers. The F6F-5N, with its Pratt & Whitney R2800-10W, had a superior performance and carried either the AN/APS 6 (for the F6F-5N) or AN/APS 4 (for the F6F-5E). Both types were operated by US Navy and Marine Corps, a total of 1,434 being produced. In addition, aircraft also went to the Royal Navy as the Hellcat II NF. While it was not the most successful nightfighter to see service, the Hellcat certainly filled a gap in the air cover of the Carrier Groups.

Hellcat II data
Powerplant: One Pratt & Whitney Double Wasp R-2800-10W. Crew: One. Length: 33ft 7in. Wingspan: 42ft 10in. Maximum speed: 380mph. Ceiling: 37,300ft. Armament: 6 x .50in machine-guns.

ticity. However, Fighter Command developed 'Trigger' as a countermeasure providing better co-ordination between the nightfighters and anti-aircraft guns. Nothing happened for the next few months, and to many, as the inevitable Allied victory approached, it seemed that the threat had evaporated.

In the late evening of 3 March 1945, Bomber Command launched almost 800 aircraft against two main targets in Germany, Kamen and the Dortmund-Ems canal, and a variety of support operations. Eight Lancasters were lost over enemy territory, most to nightfighters. To most of the crews it had seemed a fairly simple mission and, as they approached their home bases, very few thought they were in danger. The 'Trigger' plan had been suspended, and the returning bombers were routed home via the usually exclusive 'home defence' zone. However, the German nightfighter bases had been ordered to execute Gisella, and began leaving their bases at 2300, crossing the North Sea at low level to evade British radar. Having crossed the coast, the intruders climbed to the predicted return altitudes of the bombers and began searching. Some 142 Ju 88s of I/NJG2, II/NJG2, III/NJG2, III/NJG3, IV/NJG3, III/NJG4 and III/NJG5 were involved. British estimates at the time put the figure at 200 enemy aircraft.

The first attack was probably that on a No. 214 Squadron Fortress, which was attacked and damaged just after midnight and had to crash-land at Woodbridge in Suffolk. By this time the defences had been alerted, and the warning that intruders were active had been passed. Two more of No. 214 Squadron's aircraft were caught, one being shot down on approach to Oulton, and the other being damaged over Peterborough but escaping to divert to Brawdy.

During the 2½hr of the operation, the nightfighters claimed thirteen Halifaxes, nine Lancasters, one Fortress and one Mosquito shot down, plus a number of others damaged to varying degrees. While this was an undoubted success and showed that the Luftwaffe was not as 'bankrupt' as some believed, it was also a final shot, as the nightfighters had themselves suffered losses. Eight aircraft were shot down, three others apparently crashed into the ground when making low-level attacks, six crews baled out when they ran short of petrol or their damaged aircraft simply proved beyond help, and eleven aircraft were written-off or damaged while landing at their bases. Such losses could not be sustained by the nightfighter force. Aircraft were not so much of a problem, but trained crews were in very short supply. The following night, a small force of intruders, estimated at ten, returned to England but found no targets. That, in essence, was the end of the Luftwaffe intruder campaign.

The introduction of the Me 262 into night defensive operations in the latter part of 1944 had been followed by various reorganisations, including the absorbtion, in April, of Kommando Welter by 10/NJG11. The unit's primary operating base was Burg, although, on the frequent occasions when the airfield was out of use owing to bombing, the aircraft operated from a nearby autobahn. By 24 January Welter himself had scored at least four night victories in the Me 262. Then, towards the latter part of the month, the first of the two-seat Me 262Bs arrived at Staaken, near Berlin, for conversion to the nightfighter role. This involved the fitting of FuG 218, the controls and indicators being in the rear cockpit. Many of the experienced single-seater pilots were not keen on the new variant, as the radar aerials clipped 38mph off the aircraft's speed. However, very few of this type saw operational service.

German records of this period can be very vague, and the effectiveness of the Me 262 in its new role is not clear, although it is generally agreed that it was responsible for an increased loss rate among Mosquitoes operating in the Berlin area. By the end of the war Welter had been credited with twenty victories flying Me 262s. German scientists continued to develop new radars and homers; in a progress report dated 25 February, present problems and future prospects were outlined:

> Spot jamming of certain SN2 frequencies makes plotting of targets difficult and the ground jamming of Freya wavelengths also spot-jams parts of the SN2. Because of the enemy's comprehensive jamming operations, effectual use of the Lichtenstein SN2 cannot now be relied upon. Developments are under way of new frequency bands. FuG 218VR (Neptun VR) is a development of Neptun J and uses any one of six spot frequencies in the VHF band. These can be set by the operator in the air. FuG 240 (Berlin N1a) uses a new technique and new frequencies giving greater acquisition range, sharp beaming of the search lobes and a rotating lobe system to give a wider scan. It is expected to introduce the system into the Ju 88G-6 shortly.

Another type of jet aircraft that saw limited operational use in the closing stages of the war was the Arado Ar 234B, which was given the Neptun system and an underfuselage gun-pack of two 20mm cannon. These aircraft were formed into a specialist unit under Hauptmann Kurt Bonow in late March as Kommando Bonow. There is no record of any successes being scored by this unit. With airfields overrun, Allied aircraft bombing and strafing every known German airfield, and fuel and spare parts in short supply, the Luftwaffe's war was over.

As the Allies closed in on the remnants of the Nazi regime, one of the final tasks of the Allied nightfighter units was to patrol the Berlin area to try and prevent the Nazi leaders fleeing by air. There was a great fear among the Allied leadership that the German hierarchy could prolong the war if it could establish itself in a new base, a policy that Hitler had decreed some time previously with his intention of using his Bavarian 'Eagle's Nest' fortress. When the end came, in early May, all that was left was for the Allied forces to take over and inspect the remnants of the Luftwaffe. Many aircraft were taken for trials and investigation in the immediate postwar period.

The campaign in the Pacific remained intense as Allied forces island-hopped towards Japan and also sought to re-take other areas. The P-61 was the standard nightfighter in this theatre of operations, eventually equipping ten squadrons. Although the Japanese air arms continued to fight, much effort being put into kamikaze attacks in an attempt to inflict 'unacceptable losses' on Allied naval units, the bomber threat was now minimal. For the US Navy the Corsair remained the primary nightfighter type. Meanwhile, the US Marine Corps (USMC) adopted the Grumman F7F Tigercat. The two-seat F7F-3N, with its distinctive droop-nosed radome (for the SCR.720), was the most numerous of the nightfighter Tigercats. Having taken part in the support operations in Northern China in 1945, the Tigercat units later exchanged their -2Ns for -3Ns (the final variant of the Tigercat was the -4N with all-round improvements, deliveries of which took place in late 1946). By 1948 units were redesignated as all-weather (AW), and it was in this capacity that they subsequently served in Korea.

The Japanese mainland was now subjected to both day and night bombing of industries and cities. The Americans had learnt the hard way

Japanese nightfighter anti-bomber tactics

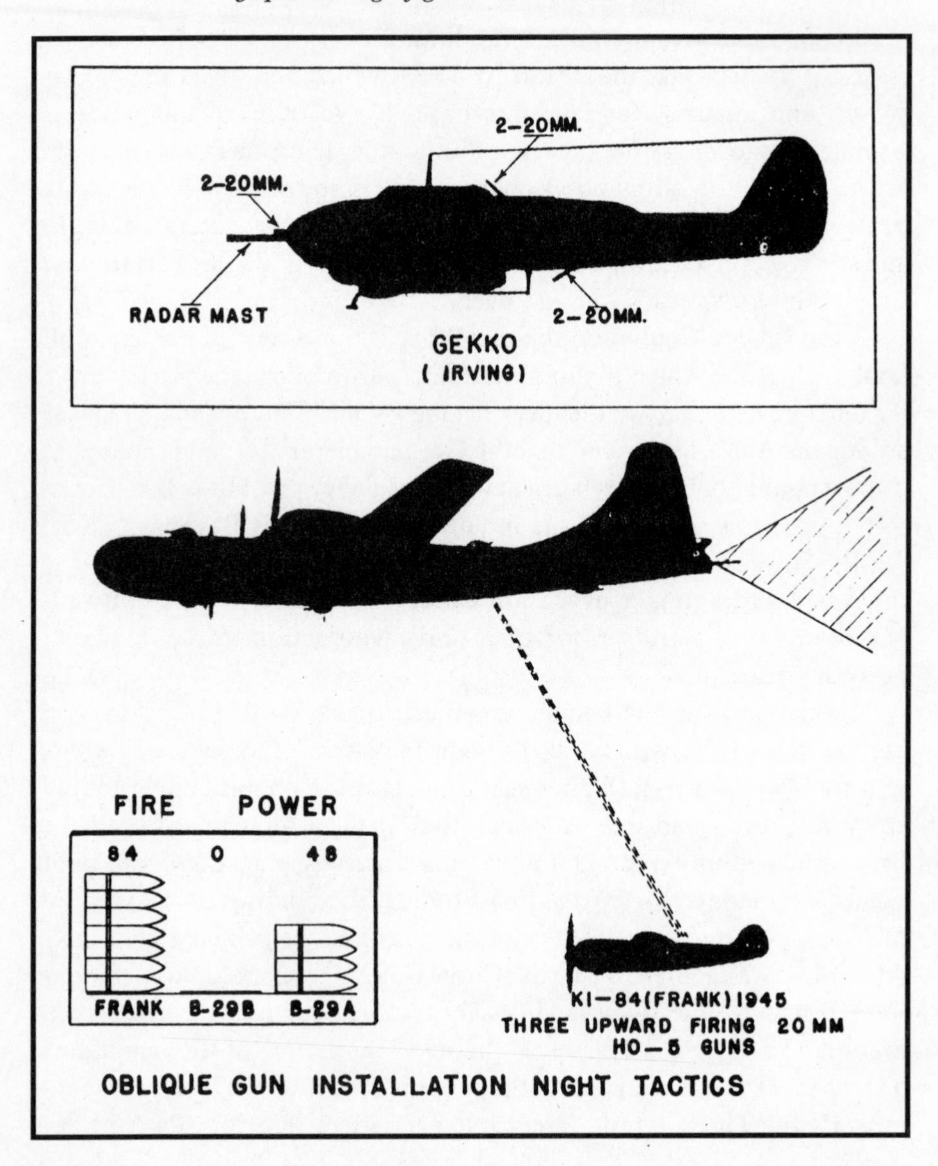

that Japanese ground forces preferred death to surrender, and the prospects for an invasion of the Japanese homeland looked grim indeed. It would inevitably be the most bloody conflict of the entire war. The bombing of Japan was a continuation of the Allied strategic bombing theory, aimed at forcing a Japanese surrender before such an invasion was required.

The B-29s roaming over Japan at night were faced by a determined but fairly ineffectual nightfighter force, the Gekko remaining the major operational type, although a limited number of Nakajima C6N1-5 Saiun (Painted Cloud) naval reconnaissance aircraft, code-named 'Myrt', saw service in the closing stages of the war, in addition to those types in use from the latter stages of 1944. The last American nightfighter kill of the war was made in August by the P-61 *Lady in the Dark* of the 548th NFS (Lieutenants Robert Clyde and Bruce Leford), who shot down a Nakajima Ki-43 Oscar near Shima. Another crew were also successful that same night.

The American heavy bomber raids were incredibly destructive, but it took the dropping of the atom bombs on Hiroshima and Nagasaki to persuade the Japanese authorities to surrender. The war with Japan, and the Second World War, ended in August 1945 after six years of bloody conflict.

Air power had formed a major element of most aspects of the conflict. The destruction wrought by the bomber forces, even before the use of atomic weapons, was on a scale greater than anything seen before. (Although the land battles and artillery bombardments of the First World War had laid waste large tracts of Europe, this was not on the universal scale of the bomber offensives.) With the bomber's ability from 1943 onwards to deliver its weapons with reasonable accuracy, even at night, it became essential to provide an efficient night defensive system, with the nightfighter as a central element. This was the scenario in which hundreds of bombers were opposed by hundreds of nightfighters and thousands of guns and searchlights; one side trying to bomb the target, the other protecting it. Then came the atom bomb. A single bomber against a single target. It would no longer be a case of trying to intercept as many bombers as possible to reduce the effectiveness of the raid. From now on the atomic bomber *had* to be found and stopped.

Chapter 9

THE JET AGE – AND MORE WARS

The ten-year period following the end of the Second World War saw very few changes in the doctrine and techniques of the nightfighter. The major change concerned aircraft types, with the introduction of jet aircraft and improved radar performance. By 1945 the Allied air forces had established all the basic elements of a night air defence system: ground-based radar units for detection and initial intercept geometry using highly-trained and skilled GCI controllers, a radar-equipped aircraft with reasonable performance and powerful weaponry to ensure a 'kill' in the often short time available for a night engagement, and a specialist crew. The major problem in the latter years of the war had been that of sorting friend from foe, despite the use of various identification systems and attempts to impose rigorous flight plan clearances. In fact, 'routeing and recognition' is still a major problem.

This time, learning from the mistakes of 1918, the western leadership sensibly prevented a total dismemberment of their military capability, although there was still a large measure of chaos as widespread disbandment, destruction and reorganisation took place to return the various countries to peacetime economies. Two elements helped ensure some measure of continuity; the need to provide an occupation force for the territory of Germany, and the growing fear of the power (and intentions) of the Soviet Union. As far as air power was concerned, the doctrine of night operations, offensive and defensive, was firmly entrenched.

The technological advantage very much lay with the British and Americans, and both continued to develop their AI capabilities. However, more important was the introduction of jet aircraft into the nightfighter role. The rationale for this was the same as it had been for day operations. Piston-engined aircraft had reached the limit of their development potential, and the only way forward, albeit with an initial backwards step in some cases, was to adopt the jet engine. This, of course, was particularly vital for fighters, as it was essential for them to have a margin of performance over the bombers. This might seem a statement of the obvious, but this need had been ignored in earlier years.

The use of the atom bomb, with aircraft as the delivery medium, had changed the balance of power and the capability of air power. A nuclear-capable air force could lay waste a country in a single day – or night. Even the requirement of accurate delivery, always a limiting factor in night and bad weather operations, had been minimised owing to the sheer explosive capability of the new weapon. It was, therefore, even more important to defend the day and night skies. The fighter had to have the performance advantage over its adversary, plus the ability to find and destroy its target as quickly as possible.

In mid-1944 the RAF had introduced the jet to day fighter operations, No. 616 Squadron being equipped with the Gloster Meteor, but it was a further seven years before the first nightfighter variant entered service, the Mosquito continuing to be the RAF's front-line nightfighter in the meantime. The initial specification, F.24/48, was issued in 1948, and the task of developing the new variant was given to Armstrong Whitworth. The basic concept was to give the aircraft a longer nose to carry the AI radar and a second seat for the radar navigator. Although initial trials took place using modified T.7 VW413, the first true Meteor nightfighter, the NF.11 prototype WA546, made its maiden flight on 31 May 1950, followed six months later by the first production aircraft, WD585. With AI in the nose, the four 20mm cannon had to be moved into the wings.

The first squadron to re-equip was No. 29 Squadron at Tangmere; this was most appropriate, as it was one of the RAF's premier nightfighter units. Powered by two Rolls-Royce Derwent 8 turbojets, the NF.11 had a maximum speed of 541mph at 10,000ft and a service ceiling of 43,000ft. The Meteor proved a capable nightfighter, and a further three variants were introduced over the next few years; the NF.12 with improved radar, the tropicalised NF.13 and, finally, the NF.14, the only one to have a clear-vision canopy over both cockpits. Final deliveries to the RAF took place in 1955, production of all variants having totalled 556 aircraft, over half of which were NF.11s. Meteor nightfighters were operated by 21 RAF squadrons, and remained operational until August 1961, although replacement by the Gloster Javelin had begun in 1956.

During the same period, de Havilland instituted a private-venture development of their Vampire day fighter to provide a jet follow-on for the superbly successful Mosquito nightfighter series. The D.H.113 prototype, given the B Conditions marking G-5-2, first flew on 28 August 1949. It was not intended for the RAF, as a preference for the Meteor had been expressed, but there was a strong export market, and the launch customer

GLOSTER METEOR

As the Meteor was the RAF's first operational jet, it was logical to adapt it for the nightfighter role. In the 1950s the series of Meteor nightfighter variants performed this role for a number of air forces. Although the Meteor first flew in March 1943 and entered service in June the following year, there was no great rush to produce a nightfighter variant. Its development was charged to Armstrong Whitworth, and the first NF.11, WA546, first flew on 31 May 1950, the type entering service the following August with No. 29 Squadron. Three other variants followed in quick succession: the NF.12 with improved radar, the tropicalised NF.13 for the Middle East, and the NF.14 with improved systems and a clear 'bubble' canopy. The RAF acquired 556 Meteor nightfighters, the NF.11 being the commonest variant. Twenty-one squadrons were equipped with Meteor nightfighters, but the era was short-lived, and units began re-equipping with the Javelin in 1956. It was not until 1961, however, that the final unit, No. 60 Squadron in Singapore, relinquished its Meteors.

Meteor NF.11 data
Powerplant: Two Rolls-Royce Derwent 8. Crew: Two. Length: 48ft 6in. Wingspan: 43ft. Maximum speed: 541mph. Ceiling: 43,000ft. Armament: 4 x 20mm cannon.

for the new type was Egypt. Then came the British ban on trade with Egypt, and suddenly the Vampires were surplus to requirements. They were duly taken on by the RAF under the designation Vampire NF.10, to speed up the jet re-equipment of the nightfighter force, the first examples going to another of the RAF's famous nightfighter squadrons, No. 25 Squadron at West Malling, in July 1951. The type eventually equipped three RAF squadrons, the total RAF production order being 78 aircraft.

Powered by a single de Havilland Goblin 3, the Vampire NF.10 had a top speed of 538mph at 20,000ft. Like the Meteor, it carried four 20mm cannon, although in this case they were mounted under the fuselage nacelle. The crew compartment followed the style of the Mosquito, with side-by-side seating. The other nightfighter to come from the de Havilland stable evolved from the Vampire's successor, the Venom. The prototype Venom NF.2, WP227, which first flew in August 1950, had a longer nose to accommodate the AI radar, plus a wider cockpit for the two crew. Its entry to service was with No. 23 Squadron in November 1953, replacing the Vampire NF.10. Earlier that same year the second of the Venom nightfighter types, the NF.3, had made its first flight, with improved radar and airframe aerodynamics and a clear canopy. The first NF.3s entered service with No. 141 Squadron in June 1955.

Thus, through the 1950s, the RAF was operating three jet nightfighters and was still the leader in the art and technology of night air defence. However, this position was abrogated in the following decade.

Top: In the on-going sequence of mosquito nightfighters, the NF.30, represented here by RK953, entered service in the latter months of the war.
Above: Junkers Ju 88G-6 3C+MN of 8/NJG4 fitted with FuG 240 'Berlin' radar. Note the nose cover in the foreground. (Manfred Griehl)
Right: A group of German nightfighter crews. (Chris Goss)

Above: The Heinkel He 219. (Manfred Griehl)

Right, top: Grumman Hellcat NF.II JX965. Some 79 Royal Navy Hellcats were equipped for the nightfighter role, the first squadron being No. 892 in April 1945.

Right, centre: In the immediate postwar period the De Havilland Mosquito remained the primary nightfighter with the Royal Air Force and a number of other air arms. Seen here is a Mosquito NF.36 with No. 39 Sqaudron in the Canal Zone, Egypt.

Right, bottom: It was only fitting that No. 29 Squadron, one of the RAF's premier nightfighter squadrons, should continue in the role after the war.

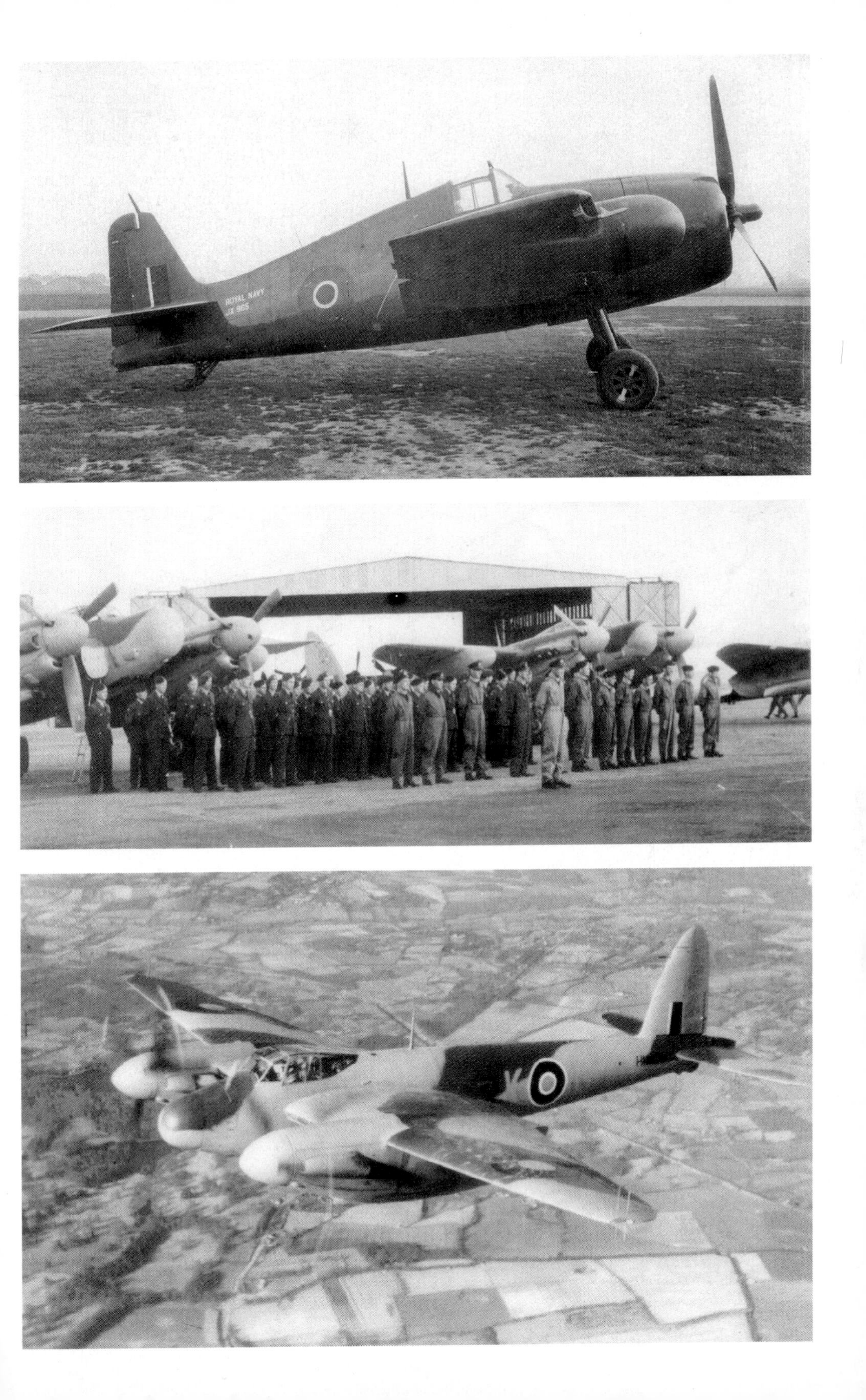
ROYAL NAVY
JX 965

Left, top: The Vought F4U-5N Corsair saw service in the Korean War, the type being the US Navy's primary nightfighter.
Left, centre: Itami Air Base, Japan, in 1950, with F4U-5N Corsair of VMF(N)-513, the Flying Nightmares'. (USMC)
Left, bottom: The one and only F4U-4N, with its AN/APS 6 radar, at the Vought works in August 1946.
Below: A Tigercat of VMF(N)-542 deployed to Korea in September 1950, where the type undertook security patrols and night ground attack missions.

Above: A Sea Hornet NF.21 in August 1946.
Right, top: After 1945 the RAF maintained a strong force of Mosquito nightfighters, a number of Auxiliary Air Force squadrons, such as No. 616 shown here with the NF.30, forming an integral part of the UK air defence network.
Right, centre: The McDonnell F2H Banshee, another US nightfighter that saw service in Korea.
Right, bottom: Northrop's F-89D Scorpion was the first purpose-designed two-seat all-weather aircraft for the USAF.

Left: The F-89J Scorpion, equipped with two MB-1 Genie nuclear (1.5KT) AAMs plus four GAR-2A Falcon AAMs; a true bomber destroyer.

Right: A Genie is loaded on to an F-89J of the 449th FIS at Ladd Air Force Base, Alaska. (Northrop)

Below: A MiG-19PM of the Soviet PVO armed with four 'Alkali' AAMs.

Above: The F3H-1N Demon carried Westinghouse APD-50 radar, giving the aircraft all-weather capability (McDonnell Douglas)
Right: An F3H Demon of VF-31, USS *Saratoga.*

Left: Meteor NF.13 WM321 of No. 219 Squadron.

WD795

Left, top: Meteor NF.13s of No. 39 Squadron deployed to Cyprus in 1956 to provide night air defence of the crowded air bases being used for the Suez operation.
Left, centre: Meteors of No. 87 Squadron in company with Javelin XA628.
Left, bottom: A superb shot of a No. 87 Squadron Meteor NF.11, piloted by Flight Lieutenant C.M. Crabb.
Above: Venom NF.2 prototype WP227.
Right, upper: The new Venom NF.2 nightfighter at Farnborough in September 1950 is inspected by John Cunningham, ex-nightfighter ace and de Havilland's chief test pilot at that time.
Right, lower: A Venom carrying two Firestreak AAMs, the RAF's first such weapon, and one that many people thought would change the nature of air combat.

Above: A trio of Venom NF.2s; WL830, WL823 and WL858 (?).
Left: A Venom of No. 253 Squadron is prepared for a night sortie during Exercise Stronghold in 1956.

Right: The control panel and display for AI Mk. 21.

Top: The Gloster Javelin, the world's first true all-weather fighter.
Above: Firestreak-armed Javelins.
Right: The era of the SAM began in the late 1950s, as these weapons were seen as key elements in any air defence system. Here a Bloodhound is under test.

Above: An F-101 Voodoo, June 1955.
Left, upper: A Voodoo with an AAM fit.
Left, lower: The navigator's console in the CF-101B, the primary day/night interceptor of the time throughout North America.

Above: The F8U-2N, the all-weather variant of the Crusader, seen here armed with four Sidewinder AAMs.
Left, upper: Gannet AEW.3 prototype XJ440, which first flew in August 1958. The advent of AEW increased the capabilities of nightfighters by adding a new dimension to the detection and control network.
Left, lower: A 1960s gathering of NATO all-weather fighters at Chaumont air base: RAF Javelin, USAF F-105 (the exception), RCAF CF-104, Belgian Air Force F-104G, French Air Force Mirage IIIC, German Air Force F-104G and Dutch Air Force F-104G.

Right, upper: Sea Vixens taxy out at Kai Tak, Hong Kong, to return to HMS *Ark Royal*. The Sea Vixen was the Royal Navy's first all-weather fighter.
Right, lower: Lockheed CF-104 Starfighters of the RCAF. The CF-104 was the Canadian version of the F-104 supersonic interceptor.

Left, upper: An impressive line-up of Lightnings of No. 74 Squadron. The Lightning was an ideal supersonic interceptor, having the speed and altitude performance to counter the Soviet bomber threat.
Left, lower: The F-4 Phantom has been one of the greatest of all fighters, seeing extensive operational service worldwide, day and night. This F-4 of No. 56 Squadron carries the standard RAF armament of four Sky Flash and four Sidewinder missiles. (BAe)

Right, upper: A long-range killer in action. An F-14 launches a Phoenix missile.
Right, lower: An F-15 of the 5th FIS. The Eagle has proved a superb interceptor and agile fighter.

Left: The F-16C cockpit. This multi-role fighter had limited night capability until the introduction of the AMRAAM system.

Above: An AIM-7 Sparrow, the standard radar AAM of the Western world, is fired by an F-16. The original F-16s in their role of day air superiority fighter were equipped with infra-red missiles. Later aircraft, with improved radar, carry not only the standard AAMs but also the advanced medium-range air-to-air missile (AMRAAM).
Left: Lockheed's stealthly F-117A reconnaissance/strike aircraft poses a new problem for any night defence system. If you cannot find it on radar, how can you shoot it down? (Lockheed)
Below: Lockheed's YF-22 combines stealth and agility for the all-weather fighter of the future. While the 'other guy' uses conventional teeth, it will be an uneven contest.

Meanwhile, the Americans were also making great technological progress and had, in many areas, overtaken the British in the development of radar and associated systems. In 1946 Major-General Curtis LeMay stated:

> The all-weather fighter is an outgrowth of the night fighter which will also be capable of operating in all types of inclement weather. It is not intended that this fighter be capable of competing under clear conditions with small interceptor type fighters.

Thus the writing was on the wall for the next generation. There would be an agile day fighter (*à la* Spitfire and Mustang), and there would be a night/bad-weather fighter (*à la* Mosquito); two distinct types of fighter, requiring different design, equipment and doctrine.

Among the aircraft put forward to meet the new all-weather requirement and replace the P-61 were the Northrop F-89 Scorpion and Curtiss XP-87 Blackhawk. Prototypes of the latter were ordered in December 1945, and initial projections looked good after the type's first flight in March 1948. The Scorpion was looked on as second favourite until the Blackhawk began to suffer development problems. This led to a rapid rethink (and the cancellation in October of the Blackhawk order), with the F-89 now having greater urgency. The type had first flown in August 1946, but it was not until July 1949 that a small production order was placed. It

DE HAVILLAND VAMPIRE AND VENOM

The Vampire was originally designed as a day fighter, but de Havilland saw prospects for an AI-equipped nightfighter variant and produced two company prototypes, the first flight taking place in August 1949. An export order was received from Egypt, and work began on the definitive nightfighter variant, the DH.113 Night Fighter, with two crew and AI Mk. X. With the sudden ban imposed on military exports to Egypt, the RAF agreed to take on the type as the NF.10. Initial deliveries were to No. 25 Squadron in July 1951.

A development of the single-seat FB.1, the Venom nightfighter variant was produced with the export market in mind. The development aircraft, G-5-3, first flew on 22 August 1950, and retained as much commonality with the standard Venom as possible. Major changes to the design were the two-seat, side-by-side cockpit and the larger nose for the radar. Initial deliveries of the NF.2 were to No. 23 Squadron in November 1953. Only two squadrons operated this variant, a further two taking the NF.2A. Development continued and the NF.3, an all-round improvement, entered service with No. 141 Squadron in June 1955. The entire project was very much an interim measure pending the production of purpose-built nightfighters; the Venom period came to an end in 1957 with the arrival of the Javelin.

Vampire NF.10 data
Powerplant: One de Havilland Goblin 3. Crew: Two. Length: 34ft 7in. Wingspan: 38ft. Maximum speed: 538mph. Armament: 4 x 20mm cannon.

appeared that the US Air Force was going to be left with a capabilities gap, and to plug this, Lockheed was asked to look at an all-weather fighter variant of its T-33 trainer, the only two-seat jet that was truly 'up and running'. As the F-94 Starfire, this made its first flight in April 1949, entering service with Continental Air Command (CONAC) squadrons at the end of the year as the F-94A. It was, as everyone knew, a somewhat thrown-together aeroplane, and there were numerous teething troubles with the weapon system, in addition to its fairly poor overall performance. Lockheed, however, was already working on major improvements, and soon came up with the F-94C, first flown in January 1950. Meanwhile, the F-89 also entered service in 1950. However, tried and tested types, such as the F4U Corsair, continued to be developed for night fighting, using improved radars.

At dawn on 25 June 1950 the forces of Communist North Korea crossed the 38th Parallel and invaded South Korea. The first major post-1945 conflict was under way. The Americans immediately promised military aid to South Korea, and within two days had secured United Nations backing. While the North Korean forces drove south, the UN forces began to assemble. Air power was to play a major part on the Allied side, the conflict becoming one of the earliest uses of the Western concept of firepower (especially aerial firepower) against low technology and numbers. Nevertheless, the Communists also employed aircraft, especially after November, when China openly joined the war.

Among the air reinforcements was the Tigercat-equipped VMF(N)-542. This unit left San Diego on 26 August 1950 on board USS *Cape Esperance*, bound for Japan, and arrived at Yokosuka on 11 September. The unit was then attached to the 5th Air Force and based at Itazuke, tasked with flying local security patrols. A week later they moved to Kimpo Air Base, near Seoul, becoming the first operational unit to be installed at the newly liberated base. As was the case with the P-61 units in the latter months of the Second World War, the nightfighter units in Korea were frequently tasked with ground-support missions, especially when the air situation offered them little 'trade'. For VMF-542 this was particularly true during the crisis around the Choisin Reservoir. This unit moved back to Japan, but continued to operate over Korea until rotated with VMF(N)-513, to whom it donated aircraft and some crews. The newly arrived unit, known as 'The Flying Nightmares' was to have more success.

November 27 1950 saw the first 'kill' by one of the USAF's strangest-looking aircraft, the F-82 Twin Mustang. The F-82, developed in 1944 in response to a requirement for a long-range escort fighter to operate over Japan, was produced by the simple expedient of joining two P-51 Mustangs together, producing a twin-fuselage aircraft. Only twenty had been delivered before the end of the war with Japan halted production. However, the delays with the production of the new jets led to renewed interest and the introduction of the P-82C (with SCR-70 radar) and P-82D (with APS-4 radar) for the nightfighter role. The radars were carried in pods under the wing centre-section linking the two fuselages, and the radar operator sat in one cockpit and the pilot in the other.

By mid-1951 the Communist air forces were employing Polikarpov Po-2s on night harassing raids against UN bases. The American troops dubbed these raids 'Bedcheck Charlie', Charlie being the term given to the Communists, and Bedcheck referring to the disturbing of their night's rest. These old biplanes proved difficult targets, because their fabric-covered structure gave a poor radar reflection (stealth!), they were painted black and they flew low and slow. However, on the night of 1 July 1951 a VMF-513 aircraft crewed by Captain E. B. Long and Warrant Officer R. C. Buckingham intercepted one of these raiders. GCI put the fighter into an intercept position, but the first pass was much too fast, as was the second. On the third set-up Long had the speed right and came in slowly enough to make a visual identification before putting in a long burst with his 20mm. The aircraft virtually disintegrated. This was the first night victory over Korea by a Marine Corps aircraft. The unit chalked up a second kill on 23 September, when Major E. A. Van Gundy and Marine Sergeant T. H. Ullom downed a Po-2 with 100 rounds of 20mm.

The first night-time jet-versus-jet combat took place on the night of 2 November 1952, when Commander Stratton, in a US Navy Douglas F3D Skyknight, destroyed a Yak-15. The Skyknight was another design of the late 1940s, produced in response to a Navy requirement for an all-weather aircraft. Grumman also submitted an aircraft for this competition, the F9F Panther, and although this lost out to the Skyknight it did enter service as a fighter-bomber. The F3D first flew in March 1948, and it soon proved to be a stable and adaptable, if not very inspiring (performance-wise) airframe. It acquired the nickname 'Willie the Whale' because of its strange shape. In Korea the Skyknight certainly proved its worth, emerging as the highest-scoring Navy fighter.

The Lockheed F-94 Starfire was also sent out to Korea, although at first its operations were confined to rear areas for fear that any loss would lead to sensitive technology being picked up by the Communists. However, in

DOUGLAS F3D SKYKNIGHT

The Skyknight was developed in response to a specification issued in August 1945 for an all-weather interceptor capable of 500mph plus, although the requirement was subsequently modified to make nightfighter capability the priority. An initial contract for three Douglas XF3D-1 prototypes was issued in April the following year, and design work began, formulating around the Westinghouse APQ-35 radar and four 20mm cannon. The dimensions of the radar scanner gave the airframe its distinct humped appearance and earned the aircraft the nickname 'Willie the Whale'.

The prototype first flew on 23 March 1948, and its success led to an order for 28 aircraft. There were few problems during the development phase, and although the aircraft did not have a startling performance it did prove to be a stable and adaptable platform. Initial deliveries of the F3D-1 were to VC-3, VC-4 and VMF(N)-542, the last also acting as a training unit. The F3D-2 was given an improved engine and a number of avionics changes; it served with a number of units in the Korean War, an aircraft of VMF(N)-513 making the first jet-versus-jet kill by downing a Yak-15 on 2 November 2 1952.

F3D-2 data

Powerplant: Two Westinghouse J34-WE-36/36A. Crew: Two. Length: 45ft 5in. Wing-span: 50ft. Maximum speed: 565mph. Ceiling: 39,400ft. Armament: 4 x 20mm cannon.

1952 the aircraft were given the task of escorting the B-29 night raids. During their period of operations the Starfire units achieved four confirmed kills, the first one, a Lavochkin La-9, being credited to an aircraft of the 319th All-Weather Fighter Squadron, crewed by Captain Ben Fithian and Lieutenant Sam Lyons, on the night of 31 January 1953. Another type to see extensive service in Korea was the McDonnell F2H Banshee. Originally, fourteen of these single-seat fighters were given extended noses to accommodate a radar. Eventually some 150 aircraft of different variants of 'Old Banjo' entered service.

Communist MiG fighters posed a threat by day and night, the night element being described by Howard Myers in this extract from his account of a reconnaissance mission by a North American RB-45:

> Having just successfully obtained target information, the all-black paint scheme on RB-45C '027 appearing to have worked in confusing the enemy searchlights, we were on our way back home with one engine shut down. Ground control told us that a flight of four fighters was on its way to escort us home – just as well, as I had seen MiGs taking off from a nearby airfield and we were no doubt the target! A few minutes later the four F-86s pitched up, two on each wing. Not long after that, ground control called two bogeys heading our way and instructed two of the fighters to peel off and engage. It later transpired that these were MiG-15s. We saw no more of the combat, and sometime later exchanged our Air Force escort for one from the Navy.

Among the most successful of the Allied nightfighters was the F4U-5N Corsair, a direct descendant of the -2N variant used in the latter part of the Second World War. The F4U-5 first flew in April 1946, and production of the nightfighter variant eventually ran to 315 aircraft, of which 101 were the 'winterised' -5NL version. As the primary Navy nightfighter, the Corsair saw service with a number of units, including VMF(N)-513, 'The Flying Nightmares', although this unit also operated other types. However, the honour of becoming the US Navy's only nightfighter ace of the Korean War went to Lieutenant Guy Bordelon of VC-3, who scored his fifth victory flying a Corsair, shooting down a Lavochkin La-11 on 16 July 1953.

June 1953 had brought a request from HQ 5th Air Force for HMS *Ocean* to form a nightfighter detachment to help counter the Po-2 attacks. The deployment of Firefly WB395 of No. 810 Squadron (Lieutenant Commander A. Bloomer and Sub-Lieutenant R. Simmonds) to Pyongtaek-ni on 7 July was intended as an operational trial of the Firefly's suitability for the role. The initial prospects looked good, so two additional aircraft arrived to start operations on 18 July, working with the GCI site at Kimpo. Aircraft flew a two-hour patrol on a racetrack combat air patrol (CAP) from Inchon to Seoul, thus giving a six-hour coverage. There was not much activity, although one mysterious aircraft was picked up and identified as a DC-3. The only 'shot in anger' was fired by Lieutenant Pete Spelling at long range against a fleeting target. It was frustrating all round; the Fleet Air Arm crews needed more experience, and the GCI site was still finding it difficult to pick up the raiders.

So it continued for the remainder of the war, as Allied air dominance took greater hold. The war eventually came to an end in July 1953, although the fragile peace has threatened ever since to re-erupt into open conflict.

During the early 1950s the Russians had also been looking at airborne radar and the introduction of jet fighters. The Mikoyan Design Bureau was the leader in this field, but the early MiG jet designs suffered from a lack of suitable engines, a partial solution being the acquisition of Western engines. A MiG-15 variant fitted with the RP-1 Izumrud (emerald) radar for gun ranging gave a limited night capability, but the first true night/all-weather variants employed by the Protivo-Vozdushnaya Oborona Strany (PVO Strany – National Air Defence Command) were the MiG-15bis/SP-1 and SP-5. However, the first production jet nightfighter was the Yakovlev Yak-25 Flashlight, a twin-engined, two-seat aircraft equipped with a nose-mounted radar (Scan Three) and a basic armament of two 37mm cannon. Entry into service was probably in 1954, and over the next few years some 1,000 aircraft were produced, some of which may have been used for reconnaissance.

These were followed by the MiG-17, still using the limited RP-1 radar, the nightfighter units being allocated 'above average' pilots because of the difficulty of their role. It was not until the MiG-17PFU/SP-6 entered service that the Soviets introduced an aircraft with an air-to-air missile (AAM) capability. In due course the MiG-19 series followed on for the same fighter roles, as well as being licence-built by a number of nations.

Military developments came thick and fast during the 1950s, given impetus by the Korean War and the growth of the so-called Cold War and the prospects of future conflict in Europe. Many of the types mentioned above continued to be improved, both with regard to performance and, more especially, in respect of their weapon systems. There were also new aircraft on the way. The US aircraft industry was particularly active, producing the likes of the McDonnell F3H Demon, Convair F-102 Delta Dagger, McDonnell F-101 Voodoo, Douglas F5D Skylancer, Convair F-106 Delta Dart and Vought F8U Crusader. All of these appeared in all-weather versions, this now being seen as the way forward. Many were single-seat, but others, like the Voodoo, were given additional specifications to become two-seat. In this case, the NF-101 made its first flight in March 1957. This aircraft was designed to carry the Falcon radar-guided AAM and the nuclear-tipped Genie missile. The development of AAMs is discussed below, as it plays a major part in the next stage of the nightfighter story. Other aircraft were designed to carry rocket pods, such as the folding fin aerial rocket (FFAR). Typical armament for certain F-89 variants might be 104 of these 2.75in Mighty Mouse rockets, the principle being the same as that used by the Luftwaffe's rocket-armed fighters in an attempt to down American heavy bombers.

The world's first true day/night all-weather fighter, the Gloster Javelin FAW.1, entered service in February 1956, joining No. 46 Squadron, RAF. This aircraft had been developed to a late 1940s requirement for a missile-armed all-weather interceptor. Although the prototype flew in November 1951, development was slow and the original variant carried a standard armament of four 30mm Aden cannon. The Javelin was produced in nine variants, the last three, FAW.7 to FAW.9, carrying Firestreak AAMs. At about the same time the Royal Navy was introducing its own all-weather fighter, the de Havilland Sea Vixen. Designed to meet the same basic

GLOSTER JAVELIN

The significance of the Gloster Javelin lay in its position as the world's first true day/night all-weather interceptor fighter equipped with AAMs. Designed to a late 1940s specification, the prototype Javelin, WD804, first flew on 26 November 1951. The original concept was for day and nightfighter variants (and a photographic reconnaissance variant) equipped with AI and four cannon, with good endurance and high performance. Development was slow, and it was not until February 1956 that the Javelin FAW.1 entered service with No. 46 Squadron, replacing that unit's Meteors. The Javelin was produced in nine variants, including one trainer type, the final three of which (the FAW.7, 8 and 9) were armed with four Firestreak AAMs in addition to the 30mm cannon. It equipped eleven RAF squadrons, replacing Meteor and Venom nightfighters, and remained in service until the late 1960s. Although no missiles were fired in anger, some squadrons were involved in operations, including Zambia in 1965 and Operation Confrontation (Indonesia) in the early 1960s. Although the Javelin carried only short-range infra-red missiles, the combination of an AI radar and AAMs was a pointer to the future. The 'all-weather' fighter had arrived, sounding the death-knell of the pure nightfighter.

Javelin FAW.9 data
Powerplant: Two Bristol Siddeley Sapphire 203/204. Crew: Two. Length: 55ft 9in. Wingspan: 52ft. Maximum speed: 620mph at 40,000ft. Ceiling: 52,000ft. Armament: Four 30mm Aden cannon in wings, and four Firestreak AAMs under wings.

requirement, although carrier-based, the prototype DH.110, WG236, flew in September 1951. Development was even slower, and the type did not join its first operational unit, No. 892 Squadron, until July 1959. It replaced the Sea Venom, and by 1961 was the Fleet Air Arm's standard all-weather fighter. Subsequent development led to the FAW.2, with overall improvements to the weapon system to enable carriage of the Red Top AAM.

So far, this account has been confined primarily to the major air forces, but it is self-evident that all air forces require a nightfighter (all-weather) capability. It was common for the types now under discussion to end up as exports or gifts to allied or friendly nations, albeit often with a degraded system capability. It would be impossible in this book to examine every aircraft type in detail, let alone every air force, so many generalisations must be made. As far as the USA was concerned, the primary purpose of many of these types was the defence of the America-Canada air environment – the need for high performance to enable them to reach and knock down Soviet nuclear bombers before they could do any damage. Throughout the analysis of the Second World War, the part played by the ground organisation – the radar network and the GCI controllers – was continually stressed. This was no less true of this period, and although significant advances had been made in aircraft radars, the role of the GCI in putting the fighter in a position from which to complete the intercept had not really changed.

As explained before, space prevents consideration of every aspect of ground-based radar development, which is a complex story. The advances that have been made in the last 30 years have seen massive increases both in detection range and tracking capability.

However, the one aspect of radar that does require a mention, owing to its increasing importance in the air defence environment, is that of the Airborne Early Warning (AEW) aircraft. In the 1990s these aircraft perform an essential role in virtually all air campaigns. The origins of AEW lie in the closing stages of the Second World War, the impetus being provided by a US Navy requirement to extend the area of coverage of ship-based radar and thus give the defenders more time to counter any threatened attack. Early development work was conducted by the Massachusetts Institute of Technology and Radiation Laboratory under Project Cadillac in early 1944. The first flight of a modified AN/APS 20 was in a stripped-down Grumman TBM Avenger in August that year. Trials went well, and production deliveries of the TBM-3W began the following spring. The RAF also tried this airborne 'GCI' in an attempt to counter the He 111s that were launching V1s. Under Operation Vapour, a radar-equipped Wellington carried an on-board controller to direct the accompanying Mosquito nightfighters.

In the postwar period, navies continued to have a requirement for such aircraft, and other types, such as the Douglas Skyraider, were modified, this type serving with a number of Western navies in the 1950s. The primary role for many of these, however, was shipping search, and it was only as technology advanced and air power doctrine developed that the potential of this type of aircraft began to be realised. By the late 1960s the concept of the Airborne Warning and Control (AWAC) aircraft had been developed to become, in effect, an airborne GCI system. This was not the only development to have a significant impact.

Although the Grumman F-14 Tomcat is now 25 years old, its crews remain the leading exponents of the long-range air-to-air kill, the combination of its Hughes AWG-9 radar and Phoenix AAMs giving it a reach of over 100 miles, day or night, fair weather or foul (but with a few limitations). In 1945 the average nightfighter had to close to within a few hundred yards to have any chance with its 20mm cannon. Air-to-air missiles have made a huge difference to all aspects of air combat, and have been responsible for the death of the pure nightfighter. Missile technology with regard to basic AAMs was well established by the latter stages of the Second World War. The German radio-guided Henschel Hs 298 was scheduled for mass production in 1945, but the war ended before anything came of the plan. This was not the only area of missile technology on which the Germans were working; the surface-to-surface V1 and V2 are well known and do not concern us here. However, they were also developing surface-to-air missiles (SAMs) such as 'Wasserfall'. These are of significance to our story, as SAMs form part of the overall defence network along with anti-aircraft guns, balloons, searchlights and nightfighters. If you can rely on a SAM/AAA network to protect your skies, you do not need fighters.

Disregarding wire-guidance, for this is not a truly practical method for air-to-air employment, there are two main types of guidance system for AAMs: radar and infra-red. In the former the aircraft radar is used to acquire and lock-on to the target to provide the missile guidance head with the information it uses to find and hit the target. There are many variations on the basic theme, but at this stage it is enough to know that the aircraft radar is the primary source of information for the missile. In infra-red guidance the seeker head in the missile looks for and homes in on infra-red energy generated by its target, the most powerful source of such emissions being the target aircraft's engines.

German technology, and scientists, formed the basis of postwar missile programmes in the USA and USSR, both of whom began extensive development work in the late 1940s. The Hughes Corporation, already well established in the field of fire-control systems and radar developments, turned its attention in 1947 to meeting a USAF requirement for a guided AAM. By the early 1950s Project Dragonfly was well under way, and deliveries of the first versions of the GAR-1 (later renamed AIM-4 Falcon) went into service with Air Defence Command in 1956. This was the first operational guided AAM in the world. However, the Russians and British were not far behind. The Russian AA-1 'Alkali' missile went into service about 1958, the same year that the de Havilland company in Britain introduced the Firestreak (previously Blue Jay) into front-line service, initial users being the Sea Venoms of No. 893 Squadron and the Javelins of No. 33 Squadron. Most of these early AAMs had restricted ranges of only a few kilometres, but even that was greater than guns could achieve. The fighter pilot no longer had to see his target; as long as the missile was locked-on, the target could be engaged, by day or night.

With the late 1950s introduction of the AA-5 'Ash', the Soviets had the largest AAM in service. Designed for the Tupolev Tu-28P long-range interceptor, it was a massive 17ft long, had a 100lb warhead, an impressive Mach 3 performance and a range of 35nm. The Tu-28P was a total departure from the concept of the small, agile fighter, being designed to be a killer of nuclear bombers. The equivalent US missile was the AIR-2A Genie developed by Douglas in the mid-1950s and given a 1.5KT nuclear warhead. Unfortunately, it had a range of only 6nm, and the pilot of the F-106 launch aircraft therefore had to fly a set attack profile, relying on the Hughes Fire Control system to track the target and provide an early lock-on in order for the him to assign and launch the missile with sufficient time to execute an escape manoeuvre. The most famous of the 1950s AAMs, and one still in service today in enhanced variants, was the AIM-7 Sparrow. Developed by Raytheon (from a Sperry Gyroscopes 'Hot Shot' programme of 1946) as a beam-riding missile, the Sparrow I entered service with the US Navy and Marine Corps in July 1956. Progress on further variants of the missile was rapid, and it established an enviable reputation. The major problem with most early AAMs was not so much the technology, although this did create support problems, as the inexperience of crews in achieving and maintaining the launch parameters.

Military conflicts have never been far away in the decades since 1945. Korea was without doubt the most significant, but there have been many others. While most of these conflicts have been limited either in scale or significance, there have been a number in which the nightfighter has played a part. The last of these in the 1950s was the Suez Crisis of 1956. Most aspects of this fall outside this study, as the Egyptian Air Force (EAF) was quickly neutralised and so played little part in countering the Anglo-French night bombing. RAF Bomber Command provided Vickers Valiants and English Electric Canberras to attack strategic targets such as airfields and barracks. The only recorded nightfighter activity by the EAF was the interception of a Valiant by a Meteor NF.13 during an attack on Almaza. The bomber crew picked up the fighter and 'climbed away as he obviously did not have the performance to stay with us'.

Meteors were also active with the Anglo-French air arm, No. 39 Squadron having moved its Meteor NF.13s to Nicosia, Cyprus, to provide night air defence of the island's crowded air bases. The threat was taken very seriously, and standing patrols were flown each night, the scale of these being reduced after the initial bombing had reduced the Egyptian capability. During November the squadron made 75 interceptions (a wide range of aircraft were ignoring routeing and recognition procedure), the biggest offenders being Lockheed Constellation airliners (22 intercepts).

The end of the Suez Crisis did not bring an end to the nightly patrols, as the squadron was now called on to help with anti-terrorist operations. For some time the Ethniki Organosis Kyprion Agoniston (EOKA) terrorist group had been seeking the union of Cyprus with Greece. An increase in terrorist action led to increased effort by the authorities to end the trouble, including an attempt to end the resupply flights that were suspected of dropping weapons to EOKA at night. Working with GCI, the Meteors were scrambled on a number of occasions, but only once was an AI contact made, the pursuit of which almost resulted in the loss of the fighter. Flight Lieutenant Derek Lewis picked up the target on his AI and directed his pilot into an intercept. They overshot the first approach and went round again. It now appeared that the target had descended very low, and the Meteor pilot positioned for a slower run when, all of a sudden, there was an almighty bang. The aircraft had skipped off the sea, but despite the damage the crew were able to return to Nicosia. It was later discovered that the target had been an Israeli Dakota on a 'training mission'.

Involvement in internal security/anti-terrorist missions was to become an increasing part of the night war in subsequent decades, although little of this involved air-to-air operations.

During the 1950s, the US Navy had recognised the need for the continued employment of a specialist fighter, and to fill this requirement the McDonnell F3H-1N Demon was developed. Equipped with Westinghouse APD-50 radar, the aircraft had limited all-weather capability, its main armament being four 20mm cannon. The first production aircraft flew in December 1953 and the type, in a number of variants, remained the US Navy's standard night/all-weather fighter until replaced by the McDonnell Douglas F-4 Phantom in the early 1960s. The original variant suffered engine problems, most of which were cured in the -2N, later models of which were also given the Hughes APGQ-51 fire control system for guns and FFAR. From 1956 onwards the FFAR element was replaced by early versions of the AAM-N-7 Sidewinder. A further variant to enter service, paving the way for future fighter development, was the -2M 'Missileer', armed with Sparrow and Sidewinder AAMs to act as the Fleet Air Defence Missile Interceptor.

The F3H-2N Demon entered service in March 1956, one of the earliest units so equipped being VF-14, the 'Top Hatters'. The first cruises were undertaken in 1956 by VF-14 aboard USS *Forrestal*, VF-124 in *Lexington* and VF-122 in *Ticonderoga*. It soon became popular with its aircrew, and by 1958 the Navy had seven operational squadrons. Also in 1958 the type was flying 'operational patrols', VF-122 flying missions in the South China Sea area as tension between Communist and Nationalist China grew. The year also saw VF-31 operational from USS *Saratoga*, flying cover missions for US forces in the Lebanon. It was a period when a host of minor conflict zones erupted, and in almost all of them American naval air power was on standby. However, the biggest crisis in the early 1960s was that over Cuba. Air bases throughout the USA were on alert, and at Key West Naval Air Station (NAS), a base well within range of Ilyushin Il-28 bombers from Cuba, aircraft were on alert day and night.

As far as development of the nightfighter itself was concerned, the emphasis remained on a two-seat AI-equipped aircraft with 20mm cannon (although AAMs were now of growing significance), working in conjunction with a ground-based GCI network. However, the 1960s were to bring a major change and the demise of the 'pure' nightfighter with the advent of the all-weather fighter.

Despite the Vought F8U Crusader being touted as an all-weather fighter, the US Navy and Marine Corps issued a requirement for a specialised nightfighter, leading to the F8U-2N (later F-8D) variant. The major change was in the avionics, the heart of the system being an AN/APQ 83 radar with enhanced capability to meet the nightfighter role. Standard armament comprised four 20mm cannon and four Sidewinder AAMs. The initial operators of the variant were VF-111, the 'Sundowners' in June 1960. The following year the F8U-2NE (later F-8E) first flew. This aircraft had even better performance, centred on the AN/APQ 94, and with the addition of two Bullpup missiles.

The Soviet Union, too, had continued development of the all-weather/nightfighter requirement, especially with regard to providing a counter to US Strategic Air Command's nuclear bombers. While various generations of MiGs, especially the MiG-21, were given such (limited) capability, designers continued to work on more specialised types. The Sukhoi Su-9 'Fishpot B' was put into production as an interim solution until it was replaced, or, in typical Soviet style, supplemented, by the 'Fishpot C', with all-round improvements and the AA-3 'Anab' missile. 1962 saw delivery of the Yak-28 'Firebar', a two-seat fighter with good radar and a range of AAM options, as well as excellent climb, altitude and speed characteristics, all essential for the interceptor of this period. Soviet doctrine continued to stress the vital role of such aircraft, and development continued with other types, such as the Su-15 'Flagon', entering service during the 1960s.

The AAM story was left at the point where three of the major air arms had just introduced the first series of missiles. Progress was rapid in the late 1950s and into the early 1960s, before, in the case of Great Britain, the financial brakes were applied. The AAM 'came of age' in the 1960s, and the requirement for a specialised nightfighter soon vanished in the face of improving technology, both in aircraft design and, more specifically, in weapon system capabilities.

The Vietnam War is important to the study of air power in a number of respects, the most significant being, first, the development of the helicopter into a major air asset, and second, the proven capability of the surface-to-air missile. It also brought the realisation that, under political constraints, air power could only achieve limited success. As outlined above, the era of the 'pure' nightfighter had now passed, and all fighters were capable of conducting operating both by day and by night (although the question of day fighter agility continued to attract attention). However, the hours of darkness still played a vital part in all military operations, especially those against what was considered to be an 'unsophisticated' enemy who relied on the night to undertake troop movements and resupply. Virtually all air offensive action during this conflict was undertaken by the US air forces, and the North Vietnamese had little in the way of an effective nightfighter defence, relying rather on an extensive network of SAM and heavy AAA. Nevertheless, the MiGs did fly night air defence missions, the MiG-21, with its superior weapon system, proving the most adept at this task.

It was during the Vietnam War that the 'package concept' of air operations was developed; a system whereby a number of specialist aircraft supported a primary mission, be it bombing or reconnaissance. The major threat to American air operations came from the SAMs and the fighters. The former were countered using 'Wild Weasel' defence suppression, and the latter by MiG CAPs. With the enemy having both a day and a night capability, it was important that the offensive aircraft had fighter protection round the clock, and to this end most of the fighter types flew night missions. Other air aspects that came into prominence during this conflict were air-to-air refuelling (AAR) and the employment of airborne control posts (ACPs). The AAR element was important because of the distances involved. Although the details are beyond the scope of this study, it is important to note that most fighter aircraft types from this period on were given an AAR capability for use day and night.

On the night of 21 February 1972 an F-4D of the 555th Tactical Fighter Squadron (TFS) scored the USAAF's first night victory of the war. Major Robert Lodge and his weapon-system operator (WSO), 1st Lieutenant Roger Locher, were on MiG Combat Air Patrol (MiGCAP) south-west of Hanoi when the Red Crown controller alerted them to MiG activity. Robert Lodge recalled the mission:

> I descended to minium en route altitude and at approximately 2123hr my WSO detected and locked on a target at the position Red Crown was calling a bandit. The target was level at zero azimuth, with the combined velocity of both aircraft in excess of 900kt. I fired three AIM-7Es, the first at approximately 11 miles, the second at 8 miles and the third at 6 miles. The first missile appeared to guide and track level, and detonated in a small explosion. The second missile guided in similar manner and detonated with another small explosion, followed immediately by a large explosion in the target area.

The F-4 remained the primary fighter throughout the Vietnam War, achieving an admirable record by day and night, although the majority of air effort was confined to the former. Other than the growth of reliance on the airborne command post (ACP – or AEW) and, after initial problems, the success of AAMs, there were no major night warfare lessons. However, two aspects showed the future path that such operations would take.

The unsatisfactory end to the Vietnam conflict, from the American point of view, provided many lessons for the air power theorists, largely orientated around the stifling requirements of political constraints, although none of these have any major significance in out particular story. If any one aspect could be said to have had a major impact on this area, it was the proven capability of the SA-2 and SA-3 SAMs as part of an integrated air defence network in the medium- to high-level band. Such radar-guided missiles work equally well day or night, so the provision of extensive belts or zones of such weapons reduces the requirement for nightfighter defence, though it does not eliminate it. Along with all other aspects of military technology, SAMs were to undergo major advances in capability, partly in response to changing threat scenarios and the countermeasures devised for aircraft employment.

The 1960s were very much a missile age, to the extent that some countries were of the opinion that the age of the manned aircraft was over; missiles would be able to achieve most major 'air' tasks. This was, of course, rapidly proved to be a facile argument, but the effect of missiles on overall military doctrine was certainly significant. In the day fighter argument it came to the point at which the designers had to reintroduce a gun to the aircraft. In the absence of a true 'dogfighting' missile it was proving impossible to make a kill once the fight was close in; on many occasions a pilot had the enemy dead to rights – if only he had a gun to fire. Air combat had certainly developed during the Vietnam War, although for the Americans this simply involved fighter versus fighter combat, the North Vietnamese having no real offensive air capability.

Throughout the 1970s and into the 1980s the F-4 Phantom was the primary air defence fighter for a great many nations. An examination of the employment of this type, therefore, provides a reasonable summary of air defence operations over two decades; a summary that in many cases still applies today, as the F-4 continues to be a front-line fighter. Over the years it has seen extensive operational employment in such diverse theatres as Vietnam and the Middle East.

The Phantom has appeared in a number of guises, the RAF and RN air defence version, the FG.1, being typical of the breed. In September 1969 No. 43 Squadron re-formed at RAF Leuchars in Scotland as part of the revision of the UK's air defence network, receiving a complement of Phantom FG.1s in the same year that the Navy had established No. 767 Squadron as its Phantom Training Unit. The UK air defence network comprised an integrated system of air defence radars (including GCI sites), SAM zones equipped with the Bloodhound missile, and a number of dedicated squadrons equipped with either the English Electric Lightning or Phantom. At the time the perceived threat was from Soviet bombers, the primary concern being their capability of launching nuclear-armed stand-off weapons. The standard armament for the Phantom was four radar-guided and four infra-red AAMs, with the option to carry a 30mm gun pod under the fuselage.

Typically, the fighters operated as a pair. Having been directed to the CAP area, the fighter pair took up a racetrack orbit orientated towards the axis from which the threat was expected. The air defence controller was monitoring the overall air situation in his given sector, and was thus in a position to allocate threats to particular weapons systems. When a raid was spotted inbound in the sector covered by the CAP, the GCI directed the formation in for an intercept. As long as the Rules of Engagement (ROE) were sound, the idea was for the fighters to make the first kills at beyond visual range (BVR), using Sparrow/Sky Flash missiles. As soon as the navigators in the Phantoms acquired the targets, they allocated missiles to those considered to be the greatest threat. With the radar kills made, the fight might well come to the 'merge' (i.e. the two formations passing each other), at which point the Phantoms would attempt to engage with the Sidewinders as the combat dissolved into a traditional dogfight. The latter scenario, of course, was not part of the nightfighter tactic; here, the radar shots would have been taken and the defenders would attempt to position for another BVR set-up.

Until the operational advent of stealth in the 1990s, nothing much changed in principle. Aircraft systems and ground systems improved as new technology entered service. Other conflicts, such as the 1973 Yom Kippur War between Israel and its Arab neighbours, confirmed many of the Vietnam lessons with regard to AAM and SAM employment. It was very much an electronic battlefield once more. This particular conflict was also used by the USSR and USA as a proving ground for new technology.

Most air arms were caught in the technology spiral, whereby more technology equalled greater cost which, under financial constraint, equalled less equipment. However, one of the great advantages of improved ground radar and the use of AEW aircraft is that fewer fighters are required in order to achieve an intercept. AEW aircraft have become a significant element of the major air forces, the Boeing E-3 AWACs being a leader in its class.

Although the foregoing was true for Western-orientated, or East–West conflict, it did not apply to all air forces. Another conflict in the early 1970s in which air power was significant was that between Pakistan and India. While the majority of air activity took place by day, both sides also flew night bombing missions against airfields, the Pakistan Air Force (PAF) Martin B-57s being particularly active. One of the few confirmed night engagements took place on 8/9 December, when an Indian Air Force fighter picked up a B-57. However, before the fighter could launch a missile the bomber had flown into the ground near Masroor while manoeuvring.

The air operations aspects of the 1982 Falklands Conflict between Argentina and Great Britain reinforced a number of air power lessons, not least of which was the need for airfields. The absence of such facilities for the British Task Force meant that they had to rely on shipborne aircraft, in this case British Aerospace Sea Harriers, for air defence. The Sea Harrier was essentially a day clear-weather aircraft, armed with short-range infra-red missiles (AIM-9 Sidewinders) and a gun. However, night air operations by the Argentinian air force were restricted to a few bombing sorties by Canberras, which were of little significance, and resupply by transport aircraft flying into Port Stanley airfield. While the latter was never a major factor in the outcome of the conflict, it was certainly something that the British would like to have stopped if they could. However, in the absence of a true radar-equipped 'nightfighter', and with the primary role of the Sea Harriers being the defence of the vulnerable task force against determined attacks by Argentine aircraft, this proved impossible.

The other Argentine air operation that proved annoying and posed a threat was the reconnaissance (tracking and targeting) by Grumman S-2E Tracker aircraft. The main Argentinian strike aircraft, Skyhawks, Mirages and Daggers, were also essentially daytime good-weather types, so in many ways the orders of battle were balanced and the night scenario was not of such critical importance. The Sea Harrier CAPs were manned dawn to dusk, and on a number of occasions the crews were operating in 'somewhat less than true daylight', although none was a real night engagement.

Throughout the 1980s, then, the basic requirement was still to prevent an enemy bomber force from reaching a position from which it could launch bomb or missile attacks against friendly territory. Almost all Western and Soviet planning and development was orientated around the Central Region of Europe, the same concepts being applicable to other major conflict areas. Thus the 24hr all-weather fighter remained the key element of the defensive system. Very few new aircraft were under development, technology being applied instead to enhancing existing airframes through the provision of improved radar and missile systems. The basic policy was that, if the airframe was reasonable, it could be kept in service through system upgrades. The major problem came with the argument as to the requirements of the fighter. The basics of the interceptor are very different from those of the air-superiority fighter, the latter implying a need to win a close-in dogfight, whereas the former uses superior technology to knock down targets before they merge (into a dogfight).

One of the perennial problems for defence planners is assessing from where the threat will come. Having done that, it is possible to make a reasonable assessment of the character of the air power that is required. With long lead times for new equipment, commonly ten years or more from concept to operational service for a major combat aircraft, this is not easy. Thus, in the Western democracies, the trend has been to acquire multi-role aircraft.

In a scenario that envisaged masses of Soviet aircraft crossing into NATO airspace in the Central Region, what was needed was a large number of fighters capable of both long-range and close-in engagements. To prevent a pre-emptive strike against crucial NATO installations, all fighter airfields kept Quick Reaction Alert (QRA)/Zulu Alert aircraft fully armed and ready to go; all part of the Cold War stand-off. While Britain maintained two fighter squadrons of F-4 Phantoms at Wildenrath in Germany, the requirements for UK defence were somewhat different. The major threat here was from long-range Soviet bombers capable of launching stand-off nuclear weapons from out over the North Sea. Thus a bomber killer was called for, a role that was performed first of all by the Lightning and then, more significantly and with greater capability, by the F-4. In both cases, air bases maintained the standard QRA arrangement. When it came to finding a replacement for these aircraft, however, most attention was devoted to such aspects as radar missile kill and long loiter time on CAP. Thus the Panavia Tornado F.2/F.3 was seen as the ideal solution.

There were many who thought that the F-4 was the ultimate fighter aircraft, but despite attempts to keep the aircraft up-to-date with changes in avionics – attempts that continued with such creations as the Israeli 'Super Phantom' – there was always going to be a limit to the type's agility and upgrade capability. While the F-4 remains a potent airframe for many countries, it has been phased out or supplemented in most NATO air arms. The next generation, the so-called 'third generation', of fighters, have all followed the standard all-weather day/night concept, though some are better than others. Each of these types is equipped with an advanced radar, in most cases with multi-mode capability, and a range of AAMs. With improvements in both radars and missiles, the air combat scenario has undergone marked changes, although there is still a perceived need for super-agility for use in dogfights. Among the types developed over the past twenty years are the Panavia Tornado F.3, Grumman F-14 Tomcat, McDonnell Douglas F-15 Eagle, McDonnell Douglas F-18 Hornet, General Dynamics F-16 Fighting Falcon, Dassault Mirage 2000, MiG-25, MiG-29, Su-27 and Saab Viggen. All of these are still operational with a number of air forces, and are likely to remain so for at least another ten years.

With the advent of the fourth-generation fighters into the Soviet inventory – the MiG-29 'Fulcrum' and Su-27 'Flanker' – the situation changed somewhat, and there was an increased need for an air superiority fighter that could handle these agile fighters. Thus was born the European Fighter Aircraft (EFA) requirement. While enhanced-agility characteristics are being examined for a number of existing fighters, such as the F-16, there has also been a general advance in missile and system capability that in part offsets the requirements for such agility. However, these considerations are beyond the scope of this study, as they really concern the use of the fighter in its day role, not its night role. It is not intended here to go into detail about any of these developments. The point being made is that fighter (nightfighter) requirements change to meet the changing threat.

When Iraq invaded Kuwait on 2 August 1990 it made the first moves in what was to become the largest military conflict since Korea, certainly in terms of air power employment. Condemnation was rapid, and under Saudi Arabian and American leadership a United Nations Coalition force was assembled and sent to the area. A 24hr air umbrella was provided by Saudi and American F-15C fighters and E-3 AWACs aircraft, the primary aim being to cover the vulnerable build-up areas from the threat of Iraqi air attack. The Iraqi air order of battle included a large number of advanced combat aircraft, including the latest MiG-29 'Fulcrum'. On paper at least it posed a significant threat by day and night. By December 1990 the military planners had devised, or in some cases dusted-off, plans for the forceful ejection of the Iraqis. Air power was to play a major part, with day and night attacks on economic and military targets. The Iraqi air defence network was well-equipped with modern systems, although it was largely based upon the Soviet style of rigid, procedural areas of responsibility. The military build-up was virtually complete by early January 1991, and included 1,800 combat aircraft. With the UN deadline rapidly approaching, the Coalition air forces made their final preparations. In the early hours of 17 January a major air attack was launched against key Iraqi targets, the intention being to disrupt the air defence system, along with the overall command and control network, as well as to target initial weapons capabilities (nuclear and chemical). The first priority of the Coalition planners was to attain air superiority, thus removing the Iraqi air threat and allowing the Coalition air forces total freedom in which to achieve either an Iraqi surrender or a favourable military situation from which to launch a ground offensive.

It is now a matter of record that the Iraqi air force elected not to contest the air war. It appears that the plan, if there truly was one, was to sit out the assault in their hardened shelters in the hope that the conflict would be terminated through political means. Allied aircraft dominated the skies in the Kuwait Theatre of Operations (KTO) day and night, the only threat being that posed by ground-based air defence systems (the extensive gun and missile batteries). Most sorties were given F-15 fighter escorts but, despite the efforts of some pilots to encourage the Iraqis to 'take part', there was no night aerial action. On the few occasions that Iraqi aircraft ventured forth in daylight, they were quickly brought to combat and destroyed or else succeeded in fleeing to Iran. Thus, the Gulf War showed that dominance of the night skies in the crucial first hours of such a conflict can be a deciding factor. The Iraqis abrogated control of their airspace (day and night). Perhaps this was due to a misplaced faith in the ability of SAMs to defend the airspace, or perhaps it was simply a desire to preserve the air arm for a later stage in the conflict. Whatever the case, both were serious errors of judgement.

One aspect of the Gulf War that does have major bearing on the future of night warfare was that of stealth technology. The term 'stealth' has been used, even over-used, for many years, but it is only with the operational employment of the Lockheed F-117 during the Gulf War that we can truly look at this aspect with any real relevance to air power doctrine. Development of this aircraft was very much aimed at countering the massive growth in radar detection capabilities and of radar-guided ground-based weapons systems, primarily SAMs but also AAA. Ever since Vietnam the Americans had appreciated the threat posed by such systems, and the advances made by the Soviet Union during the 1970s and 1980s brought the problem into sharp relief. It was considered essential to reduce, and if possible eliminate, the radar signature of the aircraft, and thus render it invisible to enemy radar. Other aircraft signatures, such as thermal, noise and infra-red, also need to be reduced or eliminated, but the Radar Cross Signature (RCS) is by far the most significant.

Thus, the F-117's technology is based upon faceting of the airframe to prevent radar energy being reflected back from the aircraft to the radar. This is not the place to discuss all aspects of the technology, such as radar-absorbent material (RAM) and the need to prevent engine intakes, turbine faces, cockpits and so on from acting as radar reflectors. In large measure the F-117 has the RCS problem, and to some extent infra-red and thermal problems, under control. The result, however, is a subsonic aircraft that is by no stretch of the imagination a manoeuvrable 'fighter'. It needs to operate at night to hide its weaknesses (or, of course, in conditions of air supremacy). The other side of the coin is being able to find and hit its own targets on a pitch-black night. Using electro-optical (EO) systems plus infra-red, combined with precision-guided munitions (PGMs), the 'Stealth Fighter' has proved to have few problems in this department.

As stealth is now considered to be an essential part of any modern combat aircraft design (although Eurofighter's designers appear to have ignored many of the basic precepts), and with an increasing appreciation that PGMs are the cost-effective way forward, the night skies are destined to be of increasing importance in any future conflict. With an effective night offensive capability on one side, it becomes essential to develop an effective night defensive capability. If ground-based systems or AWACs are unable to detect the stealth attacker, how can they direct a fighter to make an intercept? Are we looking at a return to the days of the patrol line over key targets? Even then the fighter would require an enhanced range of detection devices. At present, the high levels of stealth technology are expensive and hard to come by, and it is unlikely that there will be a major proliferation of the art for some time, at least not to the degree demonstrated by the F-117A and Northrop B-2 bomber.

Designs for future combat aircraft resemble aerial vessels (some hardly warrant the definition 'aircraft') that are more akin to the creations of science fiction than to their historical forebears. One thing is certain; day and night have become virtually the same as far as air doctrine is concerned, and the need to secure one's night skies in defence of vital installations is therefore greater than ever. When a single stealth aircraft with a PGM can destroy a key installation, be it a command centre or whatever, the need to find and destroy it takes on even greater significance. The night skies have never been more important.

APPENDICES

APPENDIX 1. GERMAN NIGHT RAIDS – FIRST WORLD WAR

The following table gives basic details of every German night raid against the UK in the First World War, by airships and aeroplanes. Information regarding targets is kept simple; although the planned targets are relatively easy to determine, the actual locations bombed tended to be more diverse, as the airships scattered bombs over a wide area. The important aspects of this table are the scale of the defence effort and the proportion of RFC to RNAS sorties.

Date	**Targets**		**Airships**	
	Plan	**Actual**	**Tasked**	**Attacked**
19/20.1.15	Humber, Thames		L3, L4, L6	L3, L4
21/22.2.15		Braintree, Colchester	1 x FF29 Seaplane	1 x FF29
14/15.4.15	Tyneside	Tyneside	L9	L9
15/16.4.15	Humber	Lowestoft, Maldon	L5, L6, L7	
29/30.4.15	E. Anglia	Ipswich, B-S-E, E Anglia	LZ38	LZ38
9/10.5.15	London	Southend	LZ38	LZ38
16/17.5.15	Kent coast	Ramsgate, Dover	LZ38	LZ38
26/27.5.15		Southend	LZ38	LZ38
31.5/1.6.15	London	London, Kent coast	LZ37, LZ38	LZ37, LZ38
4/5.6.15	London, Humber	Gravesend, Hull	L10, SL3	L10, SL3
6/7.6.15	Hull, London	London	L9, LZ37, LZ38, LZ39	L9
15/16.6.15	Tyneside		L10, L11	L10
9/10.8.15	London, Humber		L9, L10, L11, L12, L13	
12/13.8.15	London	Harwich	L9, L10, L11, L13	
17/18.8.15	London		L10, L11, L13, L14	L10
7/8.9.15	London	E. Anglia	LZ74, LZ77, SL2	
8/9.9.15	London	E. Dereham, London, E. Anglia	L9, L13, L14	
11/12.9.15	London		LZ77	LZ77
12/13.9.15	London	E. Anglia	LZ74	LZ74
13/14.9.15	London		L11, L13, L14	L13
13/14.10.15	London	London	L11, L13, L14, L15, L16	
22/23.1.16	Dover	Dover	1 x FF33b	1 x FF33b
31.1/1.2.16	Liverpool	Midlands	L11, L13, L14, L15, L16, L17, L19, L20, L21	
5/6.3.16	Rosyth, Tyne/Tees	Hull	L11, L13, L14	L11, L13, L14

Defence sorties		Effect of bombing		Remarks
RNAS	RFC	Killed	Damage	
0	2	4	£7,740	
0	0	0	?	A/c ditched and crew captured
1	0	0	£55	
2	0	0	£6,498	
4	0		£9,000	
11	0	1	£5,301	
5	0	2	£1,600	
5	0	3	£987	
15	0	7	£18,596	
5	4		£8,740	
3	0	24	£44,795	LZ37 shot down by Sub-Lt Warneford over Belgium; LZ38 bombed at Evère
5	0	18	£41,760	
21	0	17	£11,992	L12 damaged by AA guns, ditched nr Zeebrugge
4	0	7	£3,649	
6	0	10	£30,750	
3	0	18	£9,616	
7	0	26	£534,287	
2	1	0	0	
1	0	0	0	
2	0	0	£2	L13 damaged by AA guns
0	6	71	£80,020	
0	0	1	£1,591	
8	14	70	£53,832	L19 crashed in sea
1	0	18	£25,005	

31/1.4.16	London, E. Anglia		L13, L14, L15, L16, L22, L9, L11, LZ90, LZ88, LZ92	L13, L14, L15, L16, L22
1/2.4.16	N. England		L11, L17	L11, L17
2/3.4.16	London		L14, L16, L22, L13,	L14, L16, L22,
3/4.4.16	London		L11, L17	L11
5/6.4.16	N. England		L11, L13, L16	L11, L16
24/25.4.16	London	E. Anglia	L11, L13, L16, L17, L21, L23	
25/26.4.16	London		LZ87, LZ88, LZ93, LZ97, LZ26	LZ87, LZ88, LZ93, LZ97
26/27.4.16	London		LZ93	
2/3.5.16	Rosyth, Manchester	N. Yorkshire	L11, L13, L14, L16, L17, L20, L21, L23, LZ98	
19/20.5.16	Kent coast	Kent coast	7 a/c: FF33, Gotha Ursinus, Hansa-B NW	7 a/c
9/10.7.16	Dover	Dover	1 x FF33	1 x FF33
28/29.7.16	E. Anglia		L11, L13, L16, L17, L24, L31	
30/31.7.16	?	E. Anglia	?	?
31.7/1.8.16	London, E. Anglia		L11, L13, L14, L16, L17, L22, L23, L31	
2/3.8.16	E. England		L11, L13, L16, L17, L21, L31	
8/9.8.16	N.E. England	Hull	L11, L13, L14, L16, L21, L22, L24, L30, L31	
23/24.8.16	E. Anglia	E. Anglia	LZ97	LZ97
24/25.8.16	London	London (1)	L16, L21, L31, L32 (+8)	
2/3.9.16	London		L11, L13, L14, L16, L17, L21, L22, L24, L30, L32, SL8, LZ90, LZ97, LZ98, SL11	
23/24.9.16	London, Midlands	Nottingham, London	L13, L14, L16, L17, L21, L22, L23, L24, L30, L31, L32, L33	
25/26.9.16	London, Midlands	York, Bolton, Sheffield	L14, L16, L21, L22, L30, L31, L23	
1/2.10.16	London, Midlands		L14, L16, L17, L21, L24, L31, L34, L13, L22, L23, L30	
27/28.11.16	Midlands Tyneside	Midlands	L13, L14, L16, L21, L22, L24, L30, L34, L35, L36	

11	13	48	£19,431	
4	3	22	£25,568	
5	9	13	£73,113	
3	1	0	0	
1	4	1	£7,983	
19	3	1	£6,412	
6	10		£568	
1	7	0	0	
7	7+	9	£12,030	L20 crashed nr Stavanger
8	0	1	£960	
0	0	0	£48	
0	1	0	£257	
16	0			
11	1		£139	
28	4		£796	1st use of a/c carrier (HMS *Vindex*)
2	0	10	£13,196	
0	0	0	0	
9	7	9	£130,203	
6	10	4	£21,072	SL11 shot down by Lt W. Leefe Robinson (No. 39 Sqn)
14	12	40	£135,68	L33 shot down by AA guns and Brandon (No 39 Sqn)
7	8	43	£39,698	
1	14	1	£17,687	L31 shot down by 2nd Lt Tempest (No. 39 Sqn)
10	30	4	£12,482	L34 shot down by 2nd Lt Pyott (No. 36 Sqn); L21 shot down by RNAS over sea

16/17.2.17	Calais	Nr Kent?	LZ107	
16/17.3.17	London		L35, L39, L40, L41 L42	
5/6.4.17	Shipping		2 x a/c	2 x a/c
6/7.5.17	London	London	1 x Albatros	1 x Albatros
23/24.5.17	London	E. Anglia	L40, L42, L43, L44 L45, L47	
16/17.6.17	London		L42, L48, L44, L45	L42, L48
21/22.8.17	N. England		L41, L42, L44, L45, L46, L47, L35, L51	
2/3.9.17	Dover	Dover	2 x a/c	2 x a/c
3/4.9.17	Chatham	Chatham	5 x Gotha	4 x Gotha
4/5.9.17	London	London, Dover, Margate, E. Anglia	11 Gotha	9 Gotha
24/25.9.17	London	London, Dover	16 Gotha	13 Gotha
24/25.9.17	Midlands, N.E.	Rotherham, Hull	L35, L41, L42, L44, L46, L47, L50, L51, L53, L55	Minus L51
25/26.9.17	London	London, Kent coast	15 Gotha	14 Gotha
28/29.9.17	London		25 Gotha, 2 Giant	3 Gotha, 2 Giant
29/30.9.17	London		7 Gotha, 3 Giant	4 Gotha, 3 Giant
30.9/1.10.17	London, Dover		11 Gotha	10 Gotha
1/2.10.17	London		18 Gotha	12 Gotha
19/20.10.17	N.England	Birmingham, Northampton, Bedford etc	L41, L44, L45, L50, L52, L53, L54, L55	All
29/30.10.17	S.E. coast	Calais (2)	3 Gotha	1 Gotha
30/31.10.17	?	Dover	?	?
31/1.11.17	London	London etc	22 Gotha	10 Gotha (London)
5/6.12.17	London	London etc	19 Gotha, 2 Giant	16 Gotha, 2 Giant
18/19.12.17	London	London etc	15 Gotha, 1 Giant	13 Gotha,1 Giant
22/23.12.17	S.E. coast		? Gotha, 2 Giant	? Gotha, 2 Giant
28/29.1.18	London	London etc	13 Gotha, 2 Giant	7 Gotha, 1 Giant
29/30.1.18	London	London	4 Giant	3 Giant
16/17.2.18	London, Dover	London, Dover	5 Giant	4 Giant
17/18.2.18	London	London	1 Giant	1 Giant
7/8.3.18	London	London etc	6 Giant	5 Giant

1	5	0	0	
3	14		£163	
0	0		£4	
2	2	1	£510	
37	39	1	£599	
22	10	3	£28,159	L48 shot down by Lt Holder/Sgt Ashby (No. 37 Sqn)
2	19		£2,272	
0	2	1	£3,486	
0	16	132	£3,993	
0	18	19	£46,047	1 Gotha shot down by AA
0	30	21	£30,818	
4	32	0	£2,210	
2	18	0	£16,394	
0	23	0	£129	First night raid by Giant
3	30	40	£23,154	
2	35	14	£21,482	
0	19	11	£45,570	First use of new sound locator
10	68	36	£54,346	L44 shot down by AA; L45, L49, L50 crashed
0	7	0	0	
0	0	0	£2	
5	45	10	£22,822	
0	34	8	£103,408	1st Home Defence sortie by Bristol Fighter (No. 39 Sqn)
0	47	14	£238,861	Capt G. Martin Green in Camel B5192 scores first NF kill against a/c
0	18	0	0	
6	97	67	£187,350	Gotha shot down by Camel
7	73	10	£8,968	
0	60	12	£19,264	
0	69	21	£38,992	
0	42	23	£42,655	Mid-air between B.E.12 and S.E.5a

12/13.3.18	Midlands	Hull	L53, L54, L61, L62, L63	All
13/14.3.18	Midlands	Hartlepool etc	L42, L52, L56	L42
12/13.4.18	Midlands	Wigan, Coventry	L60, L61, L62, L63, L64	All
19/20.5.18	London	London etc	38 Gotha, 3 Giant	28 Gotha, 3 Giant
5/6.8.18	Midlands	Nil	L53, L56, L63, L65, L70	All

APPENDIX 2. NIGHT OPERATIONS LOG OF 1 JAGDKORPS, SEPTEMB

The following table is taken from the Operations Log of 1 Jagdkorps for the period 22 September 1943 to 12 May 1944, a period of intense activity for the German nightfighter force. The first four columns are based upon RAF records; together they provide an excellent insight into the scale and effectiveness of the nightfighter force, although the German figures only apply to 1 Jagdkorps, whereas the RAF figures reflect the total Bomber Command effort.

Date	**Primary target**	**A/c to primary target**	**A/c to other other targets**
22/23.9.43	Hannover	711	52
23/24.9.43	Mannheim	628	63
27/28.9.43	Hannover	678	64
29/30.9.43	Bochum	352	25
1/2.10.43	Hagen	243	12
2/3.10.43	Munich	294	146
3/4.10.43	Kassel	547	38
4/5.10.43	Frankfurt	406	26
7/8.10.43	Stuttgart	343	141
8/9.10.43	Hannover	504	156
18/19.10.43	Hannover	360	48
20/21.10.43	Leipzig	358	66
22/23.10.43	Kassel	569	75
3/4.11.43	Düsseldorf	589	127
10/11.43	Modane	313	29
11/12.11.43	Cannes	124	90
17/18.11.43	Ludwigshafen	83	25
18/19.11.43	Berlin/Mannheim	839	45
19/20.11.43	Leverkusen	266	44
22/23.11.43	Berlin	764	38

0	9	1	£3,474	
0	15	8	£14,280	
0	27	7	£11,673	
0	88	49	£177,317	+ 2 Rumpler C VII recce; 3 x Gotha shot down, 1 crashed
0	35	0	0	L70 shot down by D.H.4 (RNAS Great Yarmouth)

ER 1943–MAY 1944

Losses (RAF records)	NF claims		Total NF ops	NF losses	
	Destroyed	**Probables**		**Dest/msg**	**Dmgd**
26	21				
33	1	29	216	7	
39		32	217	5	
9	5	1	185	5	
2			19	1	1
9	10		193	3	
24	5	11	194	6	3
10	2	3	7		
5	3	2	39	3	
30	4	27	196	3	
18	12	1	192	2	
18	28		220	4	
45	39	8	194	6	
19	16	1	60	2	
6					
			85	6	
32	1		43		
5					
26					

23/24.11.43	Berlin	383	6
25/26.11.43	Frankfurt	262	79
26/27.11.43	Berlin/Stuttgart	443	216
2/3.12.43	Berlin	458	35
3/4.12.43	Leipzig	527	24
16/17.12.43	Berlin	483	91
20/21.12.43	Frankfurt	650	152
23/24.12.43	Berlin	379	42
29/30.12.43	Berlin	712	41
1/2.1.44	Berlin	421	58
2/3.1.44	Berlin	383	64
5/6.1.44	Stettin	348	46
14/15.1.44	Brunswick	496	135
20/21.1.44	Berlin	769	79
21/22.1.44	Magdeburg	648	195
27/28.1.44	Berlin	515	182
28/29.1.44	Berlin	677	122
30/31.1.44	Berlin	534	76
15/16.2.44	Berlin	891	179
19/20.2.44	Leipzig	823	98
20/21.2.44	Stuttgart	598	228
24/25.2.44	Schweinfurt	734	336
25/26.2.44	Augsburg	594	183
1/2.3.44	Stuttgart	557	73
2/3.3.44	Meulan-les-Meureaux	223	
7/8.3.44	Le Mans	383	
13/14.3.44	Le Mans	213	122
15/16.3.44	Stuttgart	863	253
16/17.3.44	Amiens	130	37
18/19.3.44	Frankfurt	846	200
22/23.3.44	Frankfurt	816	240
24/25.3.44	Berlin	811	212
25/26.3.44	Aulnoye	192	60
26/27.3.44	Essen	705	194
30/31.3.44	Nuremburg	795	155
5/6.4.44	Toulouse	144	61
9/10.4.44	(various)	726	

20	12	3	167	6	
13	8	1	157	6	
34	27	1	157	2	
41	43	4	139	2	
24	21	3	70	3	
25	18		92	3	
43	22	1	177	4	
17	19		166	5	
20	11	1	66	3	
28	18		167	14	
27	18	2	128	7	
16	11		143	2	
38	39	5	162	1	
35	33		96	7	1
58	37	4	169	3	3
34	19	1	167	9	
49	40	2	187	2	2
33	41	2	146	5	2
45	38	1	143	10	
79	74		294	8	
10	4		103	1	
36	24	2	209	1	2
25	16		165	7	
4	3		53		
0					
0					
2					
41	29			4	
0					
22	11		168	3	1
34	37		243	4	2
73	80	8	279	9	1
0					
9	8		105	11	1
96	101	6	246	5	3
1					
11					

10/11.4.44	(various)	908	
11/12.4.44	Aachen	341	151
18/19.4.44	(various)	1125	
20/21.4.44	Cologne	357	798
22/23.4.44	Düsseldorf	596	520
24/25.4.44	Karlsruhe	637	523
26/27.4.44	Essen	493	567
27/28.4.44	Friedrichshafen	332	639
30/1.5.44	(various)	532	
1/2.5.44	(various)	801	
3/4.5.44	Mailly-le-Camp	346	252
6/7.5.44	Mantes-la-Jolie	149	231
7/8.5.44	(various)	471	
8/9.5.44	(various)	452	
9/10.5.44	(various)	682	
10/11.5.44	(various)	618	
12/13.5.44	Louvain	120	235

APPENDIX 3. LUFTWAFFE NIGHT INTRUDER OPERATIONS

The following tables are constructed from a mixture of British and German records; many of the claims made by the intruders cannot be reconciled with RAF losses. In other instances, the details given under RAF source show the most likely candidate. There is still a great deal of research to be undertaken in respect to such material, as official records are often at variance with each other. As with all other aspects of this book, any reader who can add to the information contained in this appendix is invited to communicate with the author.

Up to 13.10.41

Date	**Time**	**NF pilot**	**Claim**
23.7.40		Fw Wiese	Wellington
23.7.40		Fw Schramm	Wellington
17.8.40	2300	Fw Laufs	Hurricane
27.8.40	2130	Ofw Merbach	Hurricane
7.9.40	2300	Oblt Hermann	Blenheim
9.9.40	2315	Fw Sommer	Blenheim
20.10.40	2130	Hpt Hülshoff	Hampden
24.10.40	2302	Fw Hahn	Wellington

19					
9	8	1	101	1	3
14					
15			165	5	
42	38	4	294	6	2
30	20	2	288	4	4
30	17	7	308	4	4
35	39	3	296	15	2
1					
9	4		41		
50					
5					
12					
12					
10					
14					
14	13	3	56		2

RAF Records

Type	**Sqn**	**Remarks**
		No record
		No record
		No record
		No record
		No record
		No record
Whitley V	58	T4170; crashed nr Cleveland Hills, 4 killed; first intruder success
Whitley V	102	P5073; P/O T Murfitt; shot down on take-off from Linton 2210hrs

24.10.40	2205	Oblt Hermann	Blenheim
24.10.40	2210	Oblt Hermann	Blenheim
28.10.40	0030	Lt Völker	Hampden
23.11.40	1830	Hpt Hülshoff	Wellington
23.11.40	1845	Oblt Hermann	Hampden
23.11.40	1840	Fw Strüning	Wellington
17.12.40	0400	Oblt Bohn	Hampden
18.12.40	0636	Ofw Beier	Hurricane
22.12.40		Lt Völker	Blenheim
22.12.40	0900	Uff Blum	Blenheim
1.1.41	1840	Lt Stradner	Wellington
2.1.41	1845	Uff Arnold	Wellington
2.1.41	1902	Fw Hahn	Whitley
3.1.41	1850	Lt Böhme	Whitley
16.1.41	0230	Oblt Schulz	Blenheim
16.1.41	0245	Oblt Schulz	Blenheim
11.2.41	0110	Oblt Schulz	Blenheim
11.2.41	0230	Hpt Jung	Wellington
11.2.41	0415	Oblt Semrau	Blenheim
11.2.41	0420	Oblt Semrau	Blenheim
11.2.41	0641	Oblt Hermann	Hampden
11.2.41	0658	Oblt Hermann	Hampden
15.2.41	1015	Fw Strüning	Hudson
15.2.41	1958	Fw Strüning	Wellington
16.2.41	0125	Oblt Bönsch	Blenheim
25.2.41	2340	Fw Ziebarth	Blenheim
27.2.41	0010	Oblt Bönsch	Blenheim
27.2.41	0140	Oblt Hermann	Blenheim
2.3.41		Oblt Bönsch	Blenheim
2.3.41	2210	Fw Mittlestädt	Blenheim
2.3.41	0115	Fw Hahn	Blenheim
13.3.41	2200	Fw Hahn	Manchester
18.3.41	0653	Fw Laufs	Wellington
18.3.41	0720	Lt Pfeiffer	Wellington
4.4.41	0100	Lt Völker	Wellington
8.4.41	2345	Fw Hahn	Hampden

Blenheim IV	17 OTU	P4858; damaged 2130hrs; landed Docking
Beaufort		Damaged 2115hrs; landed Docking
Hampden I	49	X3027; F/O J .Bufton; shot down off Skegness 0030hrs
Wellington IC	301	T2517; shot down nr Digby 2225hrs
Wellington	301	T2518; shot down on landing Digby 0135hrs.
Whitley	10	T4234; damaged 1850hrs; landed Catterick
Defiant	54 OTU	L7002; shot down Church Fenton
Defiant	54 OTU	N1542; shot down Church Fenton
Wellington IC	115	R1084; shot down W of Swaffham 0200hrs
Blenheim	21	Damaged 0350hrs; landed Bodney
Blenheim	21	Z5877; shot down over Bodney 0310hrs
Hampden I	49	AD719; shot down Langworth 0540hrs
Hampden I	144	P1164; damaged 0550hrs; landed Hemswell
Hampden I	44	P2917 (X2917); F/O Penman; damaged 2300hrs; landed Waddington
Hampden I	44	X3025; ? S/L Smales; damaged, landed Waddington
Wellington IC	218	R1009; shot down on landing Marham 2300hrs
Oxford I	2 CFS	A6107; shot down on landing Fulbeck 2308hrs 26.2
Oxford?		
Blenheim I	54 OTU	L6835; damaged 2359; landed Church Fenton
Wellington	311	P9226; damaged
Blenheim	21	R3753; damaged
Manchester I	207	L7319; shot down after take-off from Waddington
Wellington IC	149	R1474; shot down on landing Mildenhall 0610hrs
Wellington IC	221	Damaged over Gt Yarmouth
Wellington IC	115	R1470; shot down 0025hrs; crash-landed off King's Lynn
Hampden I	14 OTU	P2092; shot down Little Blytham

8.4.41	0120	Lt Feuerbaum	Hudson
8.4.41	0123	Lt Feuerbaum	Hampden
8.4.41	2250	Hpt Hülshoff	Blenheim
8.4.41	2300	Hpt Hülshoff	Hampden
8.4.41	2345	Fw Hahn	Wellington
9.4.41	2315	Uff Berschwinger	Wellington
10.4.41		Fw Laufs	Wellington
10.4.41	0130	Oblt Schulz	Wellington
10.4.41		Ofw Beier	Whitley
10.4.41		Uff Köster	?
17.4.41		Lt Völker	Hampden
17.4.41		Fw Hahn	Hampden
21.4.41		Fw Hahn	Hampden
24.4.41		Fw Giessübel	?
24/25.4.41		Lt Völker	3 x Blenheim
25/26.4.41		Fw Biehne	Blenheim
26.4.41		Lt Pfeiffer	?
30.4.41	0015	Ofw Sommer	Blenheim
30.4.41	0020	Ofw Sommer	Blenheim
30.4.41	0050	Ofw Sommer	Blenheim
30.4.41	0130	Ofw Sommer	Blenheim
3.5.41	0330	Lt Feuerbaum	Hampden
4.5.41		Ofw Hahn	Fulmar
5.5.41		Ofw Hahn	Blenheim
7.5.41		Fw Strüning	Wellington
8.5.41		Uff Köster	Wellington
8.5.41		Oblt Semrau	Wellington
8.5.41		Ofw Beier	Wellington
8.5.41		Uff Köster	Blenheim
9.5.41		Fw Strüning	Wellington
10.5.41		Oblt Harmstorf	Wellington
11.5.41		Ofw Beier	Blenheim
16.5.41		Ofw Sommer	Wellington

Hudson	206	Damaged; landed West Tofts
Whitley V	51	Z6478; damaged; landed Dishforth
Wellington IA	311	Damaged; landed E. Wretham
Wellington IC	221	R1049; shot down on take-off from Bircham Newton 0035hrs
Wellington I	11 OTU	L4253; shot down Ashwell 0040hrs
Hampden I	144	AD761; Sgt Kirby; shot down after take-off from Hemswell 0015hrs
Battle I	12 FTS	P6674; shot down landing at Harlaxton 0238hrs
Wellington IC	11 OTU	N2912; shot down over Bassingbourne 0050hrs
Blenheim I	54 OTU	L1320 crashed on landing Church Fenton; L1478 damaged landing Church Fenton; K7132 shot down nr Barnsley
Defiant I 0208hrs	54 OTU	N1568; flew into ground during combat
Whitley	77	Damaged 0314hrs; landed Topcliffe
Spitfire IIA	222	P7699; shot down nr Coltishall 0145hrs
Hurricane II	257	P3866; shot down landing at Duxford 0115hrs
Wellington	11 OTU	R3227; 0210hrs; crash-landed Wendy Village
Hampden		X3062; shot down off Mablethorpe
Oxford II	14 FTS	W6636; shot down nr Sibson 0405hrs
Wellington	103	R1397; damaged
Wellington	103	Damaged
Wellington	40	T2911; damaged
	75	R1589; damaged
Wellington IC	311	R1466; crash-landed nr Thorpe Rectory 0503hrs

18.5.41		Oblt Semrau	Blenheim
4.6.41		Ofw Beier	Blenheim
12.6.41		Uff Köster	Hampden
12.6.41		Oblt Bohn	Wellington
12.6.41		Oblt Bohn	Whitley
13.6.41		Ofw Beier	Defiant
13.6.41		Oblt Semrau	?
17.6.41		Ofw Bussmann	2 x Wellington
17.6.41		Fw Arnold	Whitley
17.6.41		Oblt Bönsch	2 x Wellington
17.6.41	0355	Ofw Sommer	Wellington
17.6.41	0405	Ofw Glas	Wellington
19.6.41		Oblt Semrau	Whitley
22.6.41	0015	Oblt Bohn	Wellington
26.6.41		Ofw Jung	Wellington
26.6.41		Oblt Bönsch	Wellington
26.6.41	0015	Oblt Bohn	Wellington
27.6.41		Fw Lüddeke	Wellington
27.6.41	2355	Ofw Sommer	Wellington
30.6.41		Fw Arnold	Wellington
30.6.41		Ofw Jung	Wellington
5.7.41		Ofw Laufs	Defiant
5.7.41		Ofw Strüning	Wellington
6.7.41		Ofw Bussmann	Wellington
6.7.41		Ofw Beier	Wellington
6.7.41		Ofw Beier	Blenheim
6.7.41		Ofw Beier	2 x Whitley
7.7.41		Fw Barth	Blenheim
7.7.41		Oblt Semrau	Blenheim
8.7.41		Fw Berschwinger	Whitley
9.7.41		Lt Schulz	Hampden
15.7.41		Fw Köster	2 x Blenheims
18.7.41		Oblt Semrau	Blenheim
18.7.41		Ofw Beier	Blenheim
19.7.41		Obl Schutze	?
22.7.41		Lt Völker	Wellington
28.7.41		Lt Bisang	?

Battle T	12 FTS	R7363; shot down 0220hrs nr Grantham
Hampden I	44	X2912; damaged over Wash
Hampden I	16 OTU	P1341; damaged 0200hrs
Anson I	16 OTU	R9691; damaged 0200hrs; landed Brackley
Anson I	16 OTU	N5014; damaged 0200hrs; landed Brackley
Wellington IC	25 OTU	R1708; shot down in circuit at Finningley
Hampden		AD788; Sgt Hind
Spitfire IIA	452	P8085; shot down landing at N. Coates 0108hrs
Blenheim IV	500	Z6041; shot down 0304hrs nr Docking
Tiger Moth II	2 EFTS	R4968; shot down nr Caxton Gibbet
Blenheim I	54 OTU	K7090; flew into ground?
Wellington IC	11 OTU	X3619; damaged 0105hrs; landed Steeple Morden
Wellington IC	11 OTU	R1334; mid-air with Ju 88 0130hrs; crashed Ashwell
Oxford II	2 FTS	V3685; shot down nr Akeman St LG

30.7.41	Fw Arnold	Whitley
8.8.41	Lt Schulz	Wellington
8.8.41	Fw Köster	Whitley
8.8.41	Ofw Beier	Blenheim
8.8.41	Ofw Beier	Halifax
8.8.41	Ofw Beier	Wellington
?.8.41	Ofw Laufs	Wellington
?.8.41	Ofw Bussmann	3 x Wellington
15.8.41	Ofw Lüddeke	Wellington
15.8.41	Ofw Lüddeke	Whitley
16.8.41	Lt Hahn	Wellington
19.8.41	Ofw Strüning	2 x Blenheim
20.8.41	Fw Köster	Blenheim
1.9.41	Fw Köster	Halifax
1.9.41	Fw Köster	Wellington
20.9.41	Oblt Semrau	?
30.9.41	Ofw Bussmann	Blenheim
3.10.41	Fw Köster	Stirling
12.10.41	Lt Hahn	Oxford
13.10.41	Ofw Strüning	B-17

From 24.8.43

Date	**RAF records**		
	Type	**Sqn**	**Remarks**
24.8.43	Lancaster BI	97	EE105; 0300. shot down over Marham
7.9.43	Stirling I	1657 CU	W7455; crash-landed Gt Thurlow
23.9.43	Lancaster BI	57	W4948; shot down in circuit E. Kirkby 0043hrs
28.9.43	Lancaster III	161	ED410; shot down nr Wickenby 0120hrs
31.3.44	Mosquito XVII	25	Damaged on taxy Coltishall
31.3.44	Lancaster III	12	LM509; damaged 0425hrs; landed at Wickenby
?.4.44	Stirling III	1654 CU	LJ450; shot down Assingham
?.4.44	B-17G	96 BG	42-97556; crashed Gt Glemham Pk 0050hrs
?.4.44	Spitfire	64	Shot down nr Skeyton
?.4.44	Mosquito	60 OTU	Shot down nr Grantham
19.4.44	Lancaster II	115	LL667; shot down on landing Witchford 0210hrs

Whitley V	10	Z6584; shot down nr Thetford 0405hrs 31.7
Wellington IC	115	T2563; shot down nr Ashmanhaugh 0220hrs
Blenheim	17 OTU	L1245; shot down Wilburton
Oxford	15 SFTS	R6156 shot doen nrTackley; W6229 crashed nr Weston-on-the Green
Wellington II	12	W5536; crashed Little Coates 0239hrs 14.8
Wellington	104	W5532; shot down nr Scunthorpe 0001hrs
	17.8	
Wellington IA	11 OTU	N3005; shot down nr Barrington 0105hrs
Wellington IC	99	R1411; shot down nr Mildenhall 0248hrs
Wellington IC	27 OTU	X9611; damaged; crash-landed nr N. Luffenham
Hampden	16 OTU	P5314; shot down nr Croughton 2310hrs
Stirling I	7	N6085; shot down Caxton Gibbet 2230hrs
Blenheim IV	51 OTU	R3617; shot down Sherrington 2117hrs

19.4.44	Lancaster I	115	LL867; shot down on landing Witchford 0348hrs
19.4.44	Lancaster I	625	ME743; shot down nr Kelstern 0230hrs
21.4.44	Master II	7 PAFU	Shot down nr Peterborough 0515hrs
22.4.44	B-24J	389 BG	44-40085; crashed on landing Hethel
22.4.44	B-24J	389 BG	42-109915; shot down nr Cantley 2217hrs
22.4.44	B-24H	448 BG	42-94744; crashed Sorlingham 2225hrs
22.4.44	B-24H	448 BG	42-73497 *Vadie Raye*; crashed in flames Seething
22.4.44	B-24H	448 BG	41-28595 *Ice Kold Katie*; overshot runway Seething
22.4.44	B-24H	448 BG	41-28240; overshot runway, hit 41-28595
22.4.44	B-24H	448 BG	42-52608; shot down nr Hopton 2207hrs
22.4.44	B-24H	448 BG	41-28843; shot down nr Kessingland 2220hrs
22.4.44	B-24H	453 BG	42-64490 *Gee Gee II*; crashed Reydon Marsh 2220hrs

22.4.44	B-24J	458 BG	42-100357; crashed 2234hrs nr Horsham St Faith
22.4.44	B-24H	458 BG	42-52353; crashed nr Lakenham 2210hrs
22.4.44	B-24H	467 BG	42-52536; crashed nr Withersdale St Mendham 2230hrs
22.4.44	B-24J	467 BG	42-52445; crashed nr Barsham 2222hrs
22.4.44	Albemarle	42 OTU	V1610; shot down nr Lowestoft
25.4.44	Lancaster III	626	DV177; crashed at Boxted 0410hrs
25.4.44	Halifax III	76	LK789; shot down nr Welney 0420hrs
27.4.44	Oxford I	18 PAFU	LX196; mid-air Me 410 circuit Church Lawford 0430hrs
23.5.44	Lancaster I	619	NN695; crashed E. Wretham 0255hrs
23.5.44	Lancaster III	582	JB417; damaged 0334hrs, landed Little Staughton
22/3.5.44	Anson I	13 OTU	LT476; damaged, landed Little Staughton
?.5.44	Stirling I	1657 CU	R9298; crashed Stradishall 0239hrs (and hit LK506 and R9283)
7.6.44	B-24H	34 BG	42-94911; crashed Wetheringsett 2325hrs
7.6.44	B-24H	34 BG	42-52738; crashed Mendlesham 2325hrs
7.6.44	B-24H	34 BG	41-29572; crashed Nedging 2350hrs
8.6.44	B-24H	34 BG	42-52695; crash-landed Eye
28.6.44	Lancaster III	90	NE145; crashed Icklingham 0226hrs
28.6.44	Lancaster I	90	LM164; damaged landing Tuddenham
28.6.44	B-24H	801 BG	42-95321; crashed Eaton Socon
Operation Gisella			
3/4.3.45	Fortress II	214	KH114; damaged 0008hrs, landed Woodbridge
3/4.3.45	Mosquito XIX	169	MM640; shot down nr Coltishall 0010hrs
3/4.3.45	Fortress III	214	HB815; crashed Oulton 0016hrs
3/4.3.45	Fortress III	214	HB802; attacked over Peterborough 0039hrs
3/4.3.45	Halifax	640	NP913; crashed nr Woodbridge 0020hrs
3/4.3.45	Halifax III	171	NA107; shot down nr S. Lopham 0025hrs
3/4.3.45	Lancaster III	12	PB476; crashed nr Alford 0029hrs
3/4.3.4	Halifax III	158	PM347; crashed nr Driffield 0030hrs
3/4.3.45	Halifax III	466	NR250; baled out Waddington 0040hrs
3/4.3.45	Halifax III	76	NA584; damaged landing Holm 0040hrs
3/4.3.45	Halifax III	158	MZ917; damaged 0051hrs; landed Lissett
3/4.3.45	Lancaster III	1654 CU	PB118; crashed Church Warsop 0057hrs
3/4.3.45	Halifax III	192	LV255; crashed Fulmodeston 0059hrs
3/4.3.45	Lancaster I	460	NG502; crashed Langworth 0100hrs

3/4.3.45	Lancaster I	1654 CU	PB708; damaged land High Ercall 0202hrs
3/4.3.45	Lancaster I	1662 CU	PD444; AH over Doncaster 0105hrs
3/4.3.45	Lancaster III	1654 CU	LM748; crashed nr Newark 0105hrs
3/4.3.45	Halifax III	1664 CU	MZ654; damaged over Dishforth 0110hrs
3/4.3.45	Halifax III	1664 CU	NA612; crashed Brafferton 0112hrs
3/4.3.45	Halifax III	466	NA179; crashed nr Elvington 0110hrs
3/4.3.45	Halifax III	347	NA235; crashed Sutton on Derwent 0115hrs
3/4.3.45	Halifax III	347	NA680; crashed nr Sleaford 0105hrs
3/4.3.45	Lancaster III	12	ME323; crashed nr Blyton 0110hrs
3/4.3.45	Lancaster III	1651 CU	ND387; crashed nr Cottesmore 0115hrs
3/4.3.45	Lancaster III	44	ME442; crashed nr Grannington 0115hrs
3/4.3.45	Lancaster I	189	NG325; dived into ground E. Rudham 0118hrs
3/4.3.45	Lancaster III	1651 CU	JB699; crashed Cottesmore 0135hrs
3/4.3.45	Halifax III	76	MZ680; crashed Cadney Brigg 0136hrs
3/4.3.45	Halifax III	10	HX322; crashed nr Knaresborough 0145hrs
3/4.3.45	Halifax III	346	NR229; crash-landed nr Croft 0209hrs
3/4.3.45	Hudson	161	Damaged nr Feltwell
3/4.3.45	Lancaster	77	Damaged
3/4.3.45	Halifax III	158	NR240; damaged; landed Middleton St George
3/4.3.45	Mosquito XX	68	NT357; crashed with engine failure Horstead Hall 0359hrs
Other dates			
?.2	Stirling IV	195	LK126; crashed Shepherd's Grove 2230hrs
17.3	Lancaster	550	NG132; crashed nr Immingham 1800hrs
20.3	Stirling IV	620	LK116; shot down nr Gt Dunmow 2145hrs
20.3	Halifax	1665 CU	Shot down nr Wittering 2145hrs

APPENDIX 4. GERMAN NIGHTFIGHTER ACES

This list consists primarily of those crew whose total score exceeded 50 kills. A number of other pilots are listed where there score is high enough at the time of their death to suggest that they would have easily passed the 50 mark had they survived a few more months. Major Hermann is included because his score of 39 was achieved solely with the 'cat's-eye' fighters. Where a pilot had day kills within the total, these are given in parentheses.

Score	Name	Unit	Remarks
110 (8)	Obst Helmut Lent	NJG 1, 2, 3	k.7.10.44
121	Maj Heinz-Wolfgang Schnaufer	NJG 1, 4	
83	Maj Prince Sayn-Wittgenstein	NJG 3, 2	k.21.2.44
71 (14)	Maj Wilhelm Herget	NJG 4, 3	
66	Obst Werner Streib	NJG 1	
65	Hpt Manfred Meurer	NJG 1, 5	k.21.1.44
64	Hpt Heinz Rokker	NJG 2	
64	Obst Gunther Radusch	NJG 1, 3, 5, 2	
64	Maj Rudolf Schonert	NJG 1, 2, 5, 100	
59	Maj Paul Zorner	NJG 2, 3, 5, 100	
58	Hpt Gerhard Raht	NJG 2	
57	Hpt Martin Becker	NJG 3, 4, 6	
56	Lt Gustave Fransci	NJG 100	
56	Hpt Heinz Struning	NJG 2,1	k.24.12.44
56	Hpt Josef Kraft	NJG 4, 5, 1, 6	
55	Hpt Heinz-Dieter Frank	NJG 1	k.27.9.43
54	Obfw Heinz Vinke	NJG 1	k.26.2.44
53	Obstlt Herbert Lutje	NJG 1, 6	
53	Hpt August Geiger	NJG 1	k.27.9.43
52	Maj Martin Drewes	NJG 1	
52	Maj Werner Hoffman	NJG 3, 5	
51	Maj Prince Lippe-Weisenfeld	NJG 2, 1, 5	k.12.3.44
50	Hpt Hermann Greiner	NJG 1	
50 (22)	Obstlt Hans-Joachim Jabs	NJG 1,50	
50 +	Lt Kurt Welter	JG 300, NJG 11	
49	St/Sgt Reinhard Kollak	NJG 1, 4	
46	Hpt Ludwig Becker	NJG 2, 1	k.26.2.43
45	Lt Rudolf Frank	NJG 3	k.26.4.44
44	Lt Paul Gildner	NJG 1	k.24.2.43
44	Hpt Reinhold Knacke	NJG 1	k.3.2.43
39	Maj Hajo Hermann	JG 300	

APPENDIX 5. ORDERS OF BATTLE – NIGHTFIGHTER UNITS IN THE SECOND WORLD WAR

ROYAL AIR FORCE

Group	Unit	Equipment	Base
3 Nov 1940:			
10	604 Sqn	Blenheim	Middle Wallop
11	23 Sqn	Blenheim	Ford
	219 Sqn	Blenheim/Beaufighter	Redhill
	141 Sqn	Defiant	Gatwick
	264 Sqn	Defiant	Rochford
	73 Sqn	Hurricane	Castle Camps
	25 Sqn	Blenheim/Beaufighter	Debden
	FIU	Blenheim/Beaufighter	Tangmere
12	29 Sqn	Blenheim	Digby, Wittering
	151 Sqn	Hurricane	Digby
	85 Sqn	Hurricane	Kirton-in-Lindsey, Caistor
13	600 Sqn	Blenheim	Catterick, Drem
Forming:			
	307 Sqn	Defiant	Kirton-in-Lindsey
	422 Flt	Hurricane	Gravesend
6 April 1941:			
9	96 Sqn	Hurricane/Defiant	Cranage
	256 Sqn	Defiant	Squire's Gate
10	87 Sqn	Hurricane	Charmy Down
	307 Sqn	Defiant	Colerne
	604 Sqn	Beaufighter	Middle Wallop
11	219 Sqn	Beaufighter	Tangmere
	23 Sqn	Blenheim/Havoc	Ford
	FIU	(various)	Ford
	264 Sqn	Defiant	Biggin Hill
	141 Sqn	Defiant	Gravesend
	85 Sqn	Hurricane/Havoc	Debden
12	151 Sqn	Defiant/Hurricane	Wittering
	25 Sqn	Beaufighter	Wittering
	29 Sqn	Beaufighter/Blenheim	Wellingore
	255 Sqn	Defiant	Kirton-in-Lindsey
13	600 Sqn	Blenheim/Beaufighter	Drem, Prestwick

Forming:			
	68 Sqn	Blenheim	Catterick
15 June 1941:			
9	96 Sqn	Defiant/Hurricane	Cranage
	256 Sqn	Defiant/Hurricane	Squire's Gate
	219 Sqn	Beaufighter	Valley
	68 Sqn	Beaufighter	High Ercall
10	600 Sqn	Beaufighter	Colerne
	87 Sqn	Hurricane	Charmy Down
	307 Sqn	Defiant	Exeter
	604 Sqn	Beaufighter	Middle Wallop
11	219 Sqn	Beaufighter	Tangmere
	23 Sqn	Havoc	Ford
	FIU	(various)	Ford
	264 Sqn	Defiant	West Malling
	29 Sqn	Beaufighter	West Malling
	85 Sqn	Havoc	Hunsdon
12	25 Sqn	Beaufighter	Wittering
	151 Sqn	Defiant/Hurricane	Wittering
	255 Sqn	Defiant/Hurricane	Hibaldstow
13	141 Sqn	Defiant	Ayr, Acklington
Forming:			
	406 Sqn	Blenheim	Acklington
25 December 1941:			
9	256 Sqn	Defiant/Hurricane	Squire's Gate
	456 Sqn	Beaufighter	Valley
	96 Sqn	Defiant/Hurricane	Wrexham
	68 Sqn	Beaufighter	High Ercall
10	125 Sqn	Defiant	Fairwood Common
	1457 Flt	Havoc T	Predannack
	247 Sqn	Hurricane	Predannack
	600 Sqn	Beaufighter	Predannack
	1454 Flt	Havoc T	Colerne
	87 Sqn	Hurricane	Colerne, Scillies
	307 Sqn	Beaufighter	Exeter
	604 Sqn	Beaufighter	Middle Wallop
	1458 Flt	Havoc T	Middle Wallop
11	219 Sqn	Beaufighter	Tangmere

	1455 Flt	Havoc T	Tangmere
	23 Sqn	Havoc/Boston	Ford
	FIU	(various)	Ford
	264 Sqn	Defiant	West Malling
	29 Sqn	Beaufighter	West Malling
	1452 Flt	Havoc T	West Malling
	1451 Flt	Havoc T	Hunsdon
	85 Sqn	Havoc	Hunsdon
12	255 Sqn	Beaufighter	Coltishall
	25 Sqn	Beaufighter	Wittering
	151 Sqn	Defiant/Hurricane	Wittering,Coltishall
	1453 Flt	Havoc T	Wittering
	1459 Flt	Havoc T	Hibaldstow
	409 Sqn	Beaufighter	Coleby Grange
13	406 Sqn	Beaufighter	Acklington
	410 Sqn	Defiant	Drem, Ouston
	141 Sqn	Beaufighter	Ayr
82	153 Sqn	Defiant	Ballyhalbert
Non-operational/forming:			
	456 Sqn	Beaufighter	Valley
	418 Sqn	Boston	Debden
	157 Sqn	Mosquito	Castle Camps
	1456 Flt	Havoc T	Honiley
	1460 Flt	Havoc T	Acklington

April 1942:

9	256 Sqn	Defiant	Squire's Gate
	96 Sqn	Defiant	Wrexham
	456 Sqn	Beaufighter	Valley
	255 Sqn	Beaufighter	High Ercall
	1456 Flt	Havoc	Honiley
10	87 Sqn	Hurricane	Charmy Down
	1454 Flt	Havoc	Charmy Down
	125 Sqn	Beaufighter	Colerne
	307 Sqn	Beaufighter	Exeter
	604 Sqn	Beaufighter	Middle Wallop
	1454 Flt	Havoc	Middle Wallop
	600 Sqn	Beaufighter	Predannack
	247 Sqn	Hurricane	Predannack
	1457 Flt	Havoc	Predannack

11	157 Sqn	Mosquito	Castle Camps
	FIU	Beaufighter	Ford
	1422 Flt	Havoc	Heston
	85 Sqn	Beaufighter	Hunsdon
	1451 Flt	Havoc	Hunsdon
	23 Sqn	Beaufighter	Manston/Tangmere
	219 Sqn	Beaufighter	Tangmere
	1455 Flt	Havoc	Tangmere
	29 Sqn	Beaufighter	West Malling
	1452 Flt	Havoc	West Malling
12	68 Sqn	Beaufighter	Coltishall
	409 Sqn	Beaufighter	Digby
	1459 Flt	Havoc	Hibaldstow
	1453 Flt	Havoc	Wittering
	151 Sqn	Defiant	Wittering
13	141 Sqn	Beaufighter	Acklington
	1460 Flt	Havoc	Acklington
	406 Sqn	Beaufighter	Ayr
	410 Sqn	Defiant	Drem
ME Cmd	89 Sqn	Beaufighter	Abu Suier

April 1943:

9	96 Sqn	Beaufighter	Honiley
	255 Sqn	Beaufighter	Honiley
	256 Sqn	Mosquito	Valley
	456 Sqn	Mosquito	Valley
10	264 Sqn	Mosquito	Colerne
	307 Sqn	Mosquito	Bolthead
	125 Sqn	Beaufighter	Fairwood Common
	406 Sqn	Beaufighter	Middle Wallop
	141 Sqn	Beaufighter	Predannack
11	23 Sqn	Mosquito	Bradwell Bay
	157 Sqn	Mosquito	Bradwell Bay
	605 Sqn	Mosquito	Castle Camps
	418 Sqn	Mosquito	Ford
	604 Sqn	Beaufighter	Ford
	85 Sqn	Mosquito	Hunsdon
	29 Sqn	Mosquito	West Malling
12	25 Sqn	Mosquito	Coltishall
	410 Sqn	Mosquito	Digby

	151 Sqn	Mosquito	Wittering
	409 Sqn	Beaufighter	Acklington
	488 Sqn	Beaufighter	Ayr
	219 Sqn	Beaufighter	Catterick
Med Air	153 Sqn	Beaufighter	Maison Blanche
Cmd	255 Sqn	Beaufighter	Setif
	600 Sqn	Beaufighter	Setif
AHQ	23 Sqn	Mosquito	Luqa
Malta	108 Sqn	Beaufighter	Luqa
W Desert	89 Sqn	Beaufighter	Castel Benito/Bersis

15 July 1944:

2	464 Sqn	Mosquito	Thorney Island
	487 Sqn	Mosquito	Thorney Island
10	68 Sqn	Mosquito	Castle Camps
	151 Sqn	Mosquito	Predannack
11	96 Sqn	Mosquito	Ford
	456 Sqn	Mosquito	Ford
	418 Sqn	Mosquito	Holmsley South
	125 Sqn	Beaufighter	Hurn
	219 Sqn	Mosquito	Bradwell Bay
	605 Sqn	Mosquito	Manston
12	307 Sqn	Mosquito	Church Fenton
	25 Sqn	Mosquito	Coltishall
85	29 Sqn	Mosquito	Hunsdon
	409 Sqn	Mosquito	Hunsdon
	264 Sqn	Mosquito	Hartford Bridge
	410 Sqn	Mosquito	Zeals
	488 Sqn	Mosquito	Zeals
	604 Sqn	Mosquito	Hurn
Med/ME	600 Sqn	Beaufighter	Follonica
	255 Sqn	Beaufighter	Foggia, Grottaglie
	256 Sqn	Mosquito	La Senia, Alghero
	153 Sqn	Beaufighter	Reghaia
	108 Sqn	Beaufighter/Mosquito	Hal Far, Catania, Alghero
	46 Sqn	Beaufighter	Edcu, Gambut, Tocra, St Jean

American squadrons:

	415th	Beaufighter	Follonica

417th	Beaufighter	Borgo, Ghisonaccia
414th	Beaufighter	Alghero
416th	Beaufighter	Pomigliano

LUFTWAFFE

1 November 1940:

I/NJG 1	Bf 110	Venlo
II/NJG 1	Bf 110	
III/NJG 1	Bf 110	Athies-Laon
I/NJG 2	Ju 88/Do 17	Gilze-Rijen
II/NJG 2	Bf 110	Deelen
I/NJG 3	Bf 110	Vechta
II/NJG 3	Bf 110	Vechta

November 1941:

I/NJG 1	Ju 88	Venlo
II/NJG 1	Bf 110	Gilze-Rijen, ?
III/NJG 1	Bf 110	Athies-Laon
I/NJG 2	Ju 88/Do 17	Catania
II/NJG 2	Do 215	Deelen
I/NJG 3	Bf 110	Sicily
II/NJG 3	Ju 88	Vechta
III/NJG 3	Ju 88	Twenthe

November 1942:

I/NJG 1	Ju 88	Venlo
II/NJG 1	Bf 110	
III/NJG 1	Bf 110	Athies-Laon
IV/NJG 1	Bf 110/Do 217	St Trond
I/NJG 2	Ju 88	Sicily/N. Africa
II/NJG 2	Ju 88	Catania
III/NJG 2		
I/NJG 3	Bf 110	
II/NJG 3	Ju 88/Do 217	Vechta
III/NJG 3	Bf 110/Ju 88	Stade
IV/NJG 3	Bf 110/Ju 88	Westerland
I/NJG 4	Bf 110/Do 217/Ju 88	Frankfurt, Florennes
II/NJG 4	Bf 110/Do 217	Frankfurt

III/NJG 4	Bf 110/Do 217	Frankfurt
I/NJG 5	Bf 110	Stendal
I/NJG 101	Bf 10110	
II/NJG 101	Bf 110	
III/NJG 101	Bf 110	

September 1943:

I/NJG 1	Ju 88/Bf 110/He 219/ Do 215B	St Trond
II/NJG 1	Ju 88/Bf 110/He 219/ Do 217	
III/NJG 1	Ju 88/Bf 110/Fw 190/ Do 215B	Twenthe
IV/NJG 1	Ju 88/Bf 110	
I/NJG 2	Ju 88/Bf 110	Gilze-Rijen
II/NJG 2	Ju 88/Bf 110/Do 215B	Melsbroek
III/NJG 2	Ju 88/Bf 110	Schiphol
IV/NJG 2	Ju 88/Bf 110	
I/NJG 3	Do 217	Vechta
II/NJG 3	Ju 88/Bf 110/Do 217	Schleswig
III/NJG 3	Ju 88/Bf 110	Stade
IV/NJG 3	Ju 88/Bf 110/Fw 190	Grove
V/NJG 3	Ju 88	
I/NJG 4	Ju 88/Bf 110/Do 217	Florennes
II/NJG 4	Bf 110	St Dizier
II/NJG 4	Bf 110	Juvincourt
IV/NJG 4	Bf 110	
I/NJG 5	Bf 110	Stendal
II/NJG 5	BNf 110/Do 217	Parchim
III/NJG 5	Bf 110/He 219/Ju 88	Werneuchen
IV/NJG 5	Bf 110/Ju 88/D0 217	
V/NJG 5	Bf 110	
I/NJG 6	Ju 88/Bf 110	
II/NJG 6	Bf 110/Ju 88	
III/NJG 6	Bf 110/Ju 88	
IV/NJG 6	Bf 110/Ju 88	Neuburg
I/NJG 7	Ju 88	Münster-Handorf

I/NJG 100	Ju 87D/Ju 88/Do 217	
II/NJG 100	Ju 88/Bf 110	
IV/NJG 100	Ju 88	
I/NJG 101	Ju 88/Bf 110/He 111/ Do 217	Schleissheim
II/NJG 101	Ju 88/Bf 110/Do 217	Lechfeld
III/NJG 101	Ju 88/Bf 110/Do 217	
I/NJG 200	Bf 110	
II/NJG 200	Fw 190	Schönfeld
III/NJG 200		Stübendorf
V/NJG 200		(various Eastern Front)
I/JG 300	Bf 109G/Fw 190	Oldenburg
I/JG 301	Bf 109G/Fw 190	Rheine
II/JG301	Bf 109G/Fw 190	
JG 302	Bf 109G/Fw 190	Döberitz, Brandis, Ludwigslust, Oldenburg

November 1944:

I/NJG 1	He 219	Münster
II/NJG 1	He 219	Düsseldorf
III/NJG 1	Ju 88G/Me 410	Fritzlar
IV/NJG 1	Ju 88G	
I/NJG 2	Ju 88G	
II/NJG 2	Ju 88G	
IV/NJG 2	Ju 88G	
I/NJG 3	Ju 88G	
II/NJG 3	Ju 88G	
III/NJG 3	Ju 88G	
IV/NJG 3	Ju 88G	
V/NJG 3	Ju 88G	Kastrup
I/NJG 4	Ju 88G	
II/NJG 4	Ju 88G	
III/NJG 4	Ju 88G	
I/NJG 5	Ju 88G/Me 410	
II/NJG 5	Ju 88G	
III/NJG 5	Ju 88G	Brandis

IV/NJG 5	Ju 88G	Krakow, Powunden
V/NJG 5	Ju 88G	Klotzsche
I/NJG 6	Ju 88G	Mainz
II/NJG 6	Ju 88G	Echterdingen
III/NJG 6	Ju 88G	
IV/NJG 6	Ju 88G	
I/NJGr 10	Ju 88G/He 219	Bonn
I/NJG 11	Bf 109G/Fw 190	Fassburg
II/NJG11	Bf 109G/Fw 190	
III/NJG 11	Bf 109G/Fw 190	Bonn
I/NJG 100	Ju 88	Prowehren, Powunden,
II/NJG 100	Ju 88	Stübendorf, Krakow,
III/NJG 100	Ju 88	Breslau, Oels,
		Hohensalz
IV/NJG 100	Ju 88	Novy Sad, Beckerek,
V/NJG 100	Ju 88	Basaid, Malacki, Novy
VI/NJG 100	Ju 88	Dvor, Weiner Neustadt,
		Steinamanger
I/NJG 101	Bf 110, Ju 88	Ingolstadt
II/NJG 101	Ju 88/Do 217	Parndorf
III/NJG 101	Ju 88/Do 217	Kitzingen
I/NJG 102	Ju 88/Do 217	Oels
II/NJG 102	Ju 88/Do 217	Schönfeld
III/NJG 102	Ju 88/Do 217	Stübendorf
JG 300	Bf 109G	
JG 301	Bf 109G	
JG 302	Bf 109G	
I/KG51	Me 410	
EKdo 410	Me 410	

UNITED STATES ARMY AIR FORCES

Unit	Formed	Disbanded
6 NFS	9.1.43	20.2.46

348 NFS	4.10.42	31.3.44
349 NFS	4.10.42	31.3.44
414 NFS	26.1.43	1.9.47
415 NFS	10.2.43	1.9.47
416 NFS		
417 NFS		
418 NFS	17.3.43	.58
419 NFS	24.3.43	20.2.47
420 NFS	25.3.43	31.3.44
421 NFS	30.4.43	20.2.47
422 NFS	14.7.43	30.9.45
423 NFS		
424 NFS	24.11.43	31.3.44
425 NFS	23.11.43	25.8.47
426 NFS	8.12.43	5.11.45
427 NFS	19.1.44	29.10.45
547 NFS	18.2.44	20.2.46
548 NFS	23.3.44	19.12.45
549 NFS	19.4.44	19.2.47
550 NFS	3.5.44	4.1.46

ROYAL NAVY

Unit	Equipment	Formed	Disbanded
746 Sqn	Fulmar, Firefly, Hellcat	23.11.42	30.1.46
784 Sqn	Fulmar, Firefly, Hellcat	1.6.42	.9.46
891 Sqn	Hellcat	1.6.45	24.9.45
892 Sqn	Hellcat	1.4.45	19.4.46
1790 Sqn	Firefly	1.1.45	3.6.46
1791 Sqn	Firefly	15.3.45	23.9.45
1792 Sqn	Firefly	15.5.45	17.4.46

INDEX